SIXTEENTH SYMPOSIUM ON NUCLEIC ACIDS CHEMISTRY

HELD AT
NARASHINO, JAPAN
OCTOBER 5th – 7th, 1989

Symposium Organizer:

Professor Hiroshi Takaku
Department of Industrial Chemistry
Chiba Institute of Technology
Narashino, Chiba 275
Japan

NUCLEIC ACIDS SYMPOSIUM SERIES

No.21

IRL PRESS
at
OXFORD UNIVERSITY PRESS
Oxford New York Tokyo

IRL Press
Eynsham
Oxford
England

First published 1989

British Library Cataloguing in Publication Data

Symposium on Nucleic Acids Chemistry (*16th:1989: Naraschino, Japan*)
Sixteenth Symposium on Nucleic Acids Chemistry
1. Nucleic acids.
I. Title II. Takaku, Hiroshi III. Series
547.7′9

ISBN 0 19 963127 1

Printed by Information Press Ltd, Oxford, England.

Contents

Contents continued overleaf

Contents continued overleaf

Alkylated cAMP derivatives: selective synthesis and biological activities

Shigehiro Kataoka, Junko Imai, Nobuyuki Yamaji, Motohiko Kato, Makoto Saito[1], Tomie Kawada[2], and Shoichi Imai[2]

Research and Development Division, Kikkoman Corporation, Noda-shi, Chiba 278, [1]Noda Institute for Scientific Research, Noda-shi, Chiba 278, and [2]Department of Pharmacology, University School of Medicine, Niigata-shi, Niigata 951, Japan

ABSTRACT

Many alkylated cAMPs have been prepared and tested for their antitumor and cardiac activities. Treatment of cAMP with several bases and alkyl bromides gave alkyltriesters of cAMP (2), N^6,N^6,2'-O-trialkyl cAMPs (3), N^6,2'-O-dialkyl cAMPs (4) and 2'-O-monoalkyl cAMPs (5) in one step. N^6,N^6-dialkyl cAMPs (6) were prepared from 2'-O-tosyl cAMP by the similar alkylation, followed by detosylation. Synthesis of N^6-monoalkyl cAMPs (7) was achieved by the reductive alkylation of cAMP with aldehydes in one step. Alkyl triesters of cAMP exhibited antitumor activities against P815 cells. N^6-mono and N^6,N^6-dialkyl cAMPs showed significant cardiac activities.

INTRODUCTION

Adenosine 3',5'-cyclic phosphate (cAMP,1) is related to many biological reactions as a second messenger of various hormones.[1] Numerous derivatives of cAMP have been synthesized to obtain substances that might have a better biological action than the original nucleotide. In this paper, we wish to report the synthesis of new cAMP derivatives with selective alkylation of 1 and their antitumor and cardiac activities.

RESULTS AND DISCUSSION

In the alkylation of the phosphate moiety, 1 was treated with alkyl bromides in DMA at 75 - 110°C in the presence of tributylamine to afford alkyltriesters of cAMP (2, R_4=ethyl - cetyl) in 14 - 79% yields.[2] For the purpose of the alkylation of the base and/or sugar moieties, 1 was treated with alkyl bromides and NaH or tert-BuOK in DMSO to afford N^6,N^6,2'-O-trialkyl cAMPs (3, R_1=ethyl - myristyl and benzyl) in 24 - 58% yields. And also this alkylation was carried out with MeONa as a base to afford N^6,2'-O-dialkyl cAMPs (4, R_1=propyl - heptyl) in 17 - 32% yields. Synthesis of 2'-O-monoalkyl cAMPs (5, R_3=propyl - myristyl and benzyl) was achieved by the treatment of 1 with alkyl bromides in aqueous KOH-dioxane solution in the presence of 18-crown-6 in 11 - 51% yields. Then N^6,2'-O-dibenzyl cAMP was prepared by the similar procedure for the synthesis of 5 in 41% yield. N^6,N^6-dialkyl cAMPs (6, R_1=methyl - hexyl and benzyl) were synthesized from 2'-O-tosyl cAMP by the same method as the synthesis of 3, followed by detosylation in aqueous NaOH solution in 37 - 70% overall yields. Synthesis of N^6-monoalkyl cAMPs (7, R_1=ethyl - myristyl, benzyl and furfuryl) was accomplished by the reductive alkylation of 1 using aldehydes and sodium cyanoborohydride in acetic acid in 39 - 74% yields.[3] This reductive alkylation was applied to the preparation of N^6-alkyl-8-benzylthio cAMPs(8, R_1=propyl - myristyl, benzyl, furfuryl and phenetyl), since 8-benzylthio-cAMP was known to be a potent cardiotonic agent.[4]

These compounds were tested for antitumor and cardiac activities. Compound 2 (R_4=nonyl,decyl,dodecyl and lauryl) showed significant antitumor activities against P815 cells in vitro and in vivo (a total packed cell volume method). In cardiac activity N^6-heptyl and N^6,N^6-dipentyl cAMPs exhibited positive inotropic and weak negative chronotropic effcts on the isolated guinea pig papillary muscles and right atria; the results thus obtained will be discussed.

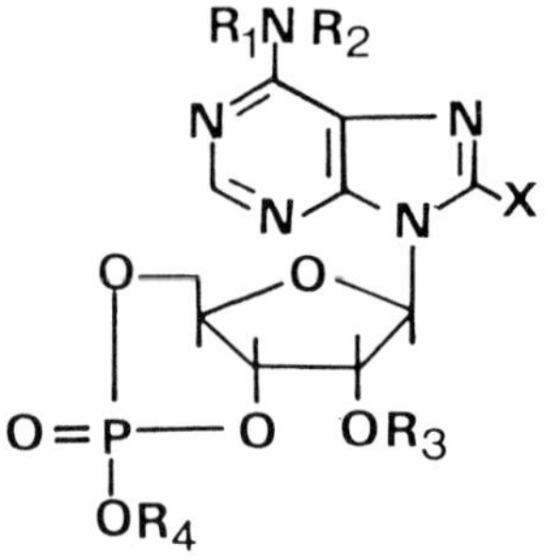

compd.	R_1	R_2	R_3	R_4	X
2	H	H	H	Al*	H
3	Al	Al	Al	H	H
4	Al	H	Al	H	H
5	H	H	Al	H	H
6	Al	Al	H	H	H
7	Al	H	H	H	H
8	Al	H	H	H	SBn

*Al=alkyl, see the texst.
SBn=SCH_2Ph

REFERENCES

1. Drummond,G.I., Greengard,P., and Robison,G.A. (1975) "Advances in cyclic Nucleotide Research," Vol.5, Raven Press, New York, pp.265-279.
2. Kataoka,S., Isono,J., Yamaji,N., and Kato,M. (1986) Chem. Lett., 1221.
3. Kataoka,S., Isono,J., Yamaji,N., Kato,M., Kawada,T., and Imai,S. (1988) Chem. Pharm. Bull., 36, 2212.
4. Yamaji,N., Kato,M., Matsubara,I., Shimamoto,N., Miura, K., and Imai, S. (1980) Chem. Pharm. Bull., 28, 1683.

Enzyme catalyzed acylation and deacylation of the sugar moieties of nucleosides

Kenji Nozaki, Atsuhiko Uemura, Jun-ichi Yamashita, and Mitsugi Yasumoto

Chemical Synthesis Research Laboratory, TAIHO Pharmaceutical Co., Ltd., Kodama-gun, Saitama 367-02, Japan

ABSTRACT

We have found that a lipase from *Pseudomonas fluorescens* (PFL) accelerated regioselective acylation of 2'-deoxynucleosides with the use of acid anhydrides as acyl donor in dry polar solvents. Different regioselective deacylation of 3',5'-di-O-acyl-2'-deoxynucleosides was found to take place when a lipase (PFL) or a protease from *Bacillus subtilis* (Subtilisin) was used.

INTRODUCTION

Regioselective acylation and deacylation of sugars and related compounds mediated by enzymes have been developed as a useful synthetic method in organic reaction.[1-4] However, there is no report on regioselective acylation and deacylation of 2'-deoxynucleosides. We now wish to report a simple and useful method for enzymatic acylation of 2'-deoxynucleosides by use of an acid anhydride. And enzymatic regioselective deacylation of 3',5'-di-O-acyl-2'-deoxynucleosides with a PFL and Subtilisin, respectively.

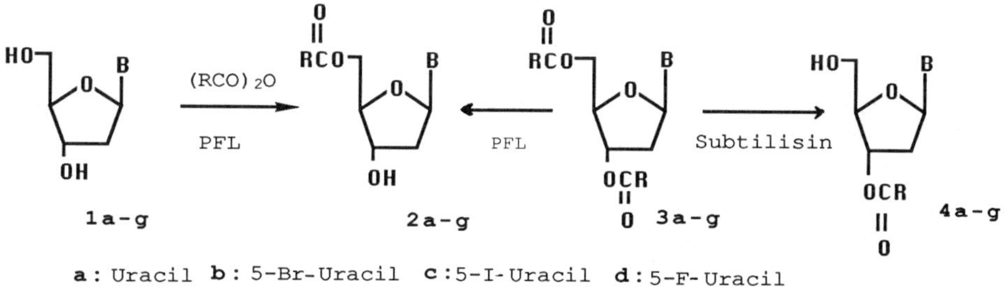

a: Uracil **b**: 5-Br-Uracil **c**: 5-I-Uracil **d**: 5-F-Uracil
e: 5-Trifluoromethyl-Uracil **f**: Thymine **g**: Adenine

ENZYMATIC ACYLATION

2'-Deoxy-5'-O-hexanoyluridine (**2-a**): Crude PFL (100 mg) was added to a solution of **1-a** (1 mmol) and hexanoic anhydride (3 mmol) in dry DMA (3 mmol) at room temperature,and the mixture was vigorously stirred for 24 hr. The reaction was quenched by filtering off the enzyme. The enzyme portion was washed with 25 ml of ethyl acetate. After standard work-up and column chromatography on silica gel gave **2-a** in 70 % yield.

ENZYMATIC DEACYLATION

3-a (0.1 mmol) in DMF(0.2 ml) and PFL(50 mg) were added to 0.1 N Phosphate buffer (1.6 ml;pH 7), then the mixture was stirred at 25 ^{o}C After 18 hr, the products were isolated by extracting the reaction mixture with ethyl acetate. After standard work-up and column chromatography on silica gel gave **2-a** in 71 % yield. The same procedure with subtilisin affords **4-a** in 22 % yield.

The presents reactions proved a method for regioselective acylation and deacylation at the 3'-or 5'- position of 2'-deoxy nucleosides.

ACKNOWLEDGEMENT

We thank Amano Seiyaku Co.,Ltd. for the supply of enzymes.

REFERENCES

1)Therisod, M; Klivanov, A. M.: *J. Am. Chem. Soc.*,**1986**,*108*,5638.
2)Sweers, H. M.;Wong, C.-H.: *J. Am. Chem. Soc.*,**1986**,*108*,6421.
3)Riva,S.;Chopineau, J.;Kieboom,A.P.G;Klivanov,A.M.: *J. Am. Chem. Soc.*,**1988**,*110*,584.
4)Hennen, W. J.;Sweers,H. M.;Wang, Y.-F.;Wong, C.-H.: *J. Org. Chem.*,**1988**,*53*,4939.

Oxidation of pyrimidine bases and their derivatives by sodium peroxodisulfate

Toshio Itahara, Naoko Ide, and Akihiro Nishino

Institute of Chemistry, College of Liberal Arts, Kagoshima University, Korimoto, Kagoshima 890, Japan

ABSTRACT

Treatment of 1,3-dimethyluracil (1) with sodium peroxodisulfate in water at 80 °C under nitrogen gave 5-hydroxy-1,3-dimethyluracil (6) as a main product. Under similar conditions, oxidation of 5-fluoro-1,3-dimethyluracil (2) gave a 6,6'-dimeric compound (10) together with other products. Similar oxidation of 5-bromo-1,3-dimethyluracil (3) and 1,3,6-trimethyluracil (4) was also investigated. Furthermore, reaction of 1,3-dimethylthymine (5) and adenine in the presence of sodium peroxodisulfate gave coupling products (12) and (13).

INTRODUCTION

Recently we reported that reaction of thymines with $Na_2S_2O_8$ in water at 80 °C under nitrogen resulted in the selective oxidation of the 5-methyl group of thymines via thymine cation radicals.[1] We now report an oxidation of uracils by $Na_2S_2O_8$ in water and coupling between 1,3-dimethylthymine and adenine.

RESULTS AND DISCUSSION

A solution of 1,3-dimethyluracil (1) and $Na_2S_2O_8$ in water was heated at 80 °C under nitrogen atmosphere. The reaction mixture was separated by liquid chromatography with low pressure pump to yield (6), (7), (8), and (9). Under similar conditions, oxidation of 5-fluoro- and 5-bromo-1,3-dimethyluracils (2) and (3) and 1,3,6-trimethyluracil (4) was further investigated. The results are summarized in Table 1. The reaction may be explained as Scheme 1. The results of the reaction of (1) and (2) suggest that water preferentially attacks position 5 of uracils. Oxidation of uracil (3 mmol) by $Na_2S_2O_8$ (6 mmol) also gave 5-hydroxyuracil (0.43 mmol).

Table 1. Oxidation of 1,3-Dimethyluracils by $Na_2S_2O_8$

Substrate (mmol)	$Na_2S_2O_8$ mmol	Water ml	React. time/ h	Conv. %	Product Isolated Yield (%)[a]
1 (1)	2	50	5	47	6(32), 7(23), 8(4), 9(4)
1 (1)	2	100	5	58	6(50), 7(22), 8(3), 9(3)
2 (1)	1	50	5	54	8(7), 9(19), 10(19)
3 (1)	2	100	3	57	8(23), 9(23)
4 (1)	1	50	5	15	11(80)
6 (0.5)	0.5	50	5	52	8(70), 9(4)

[a]Yield based on the substrate consumed.

Scheme 1

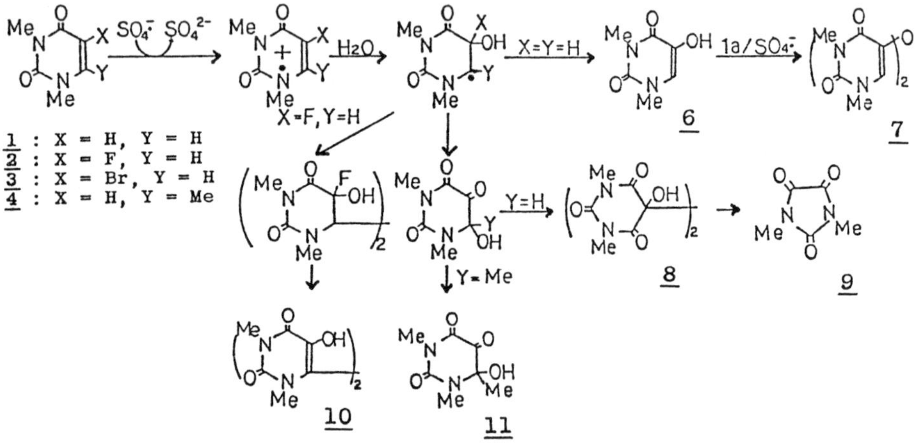

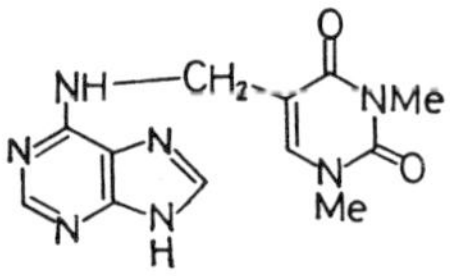

12

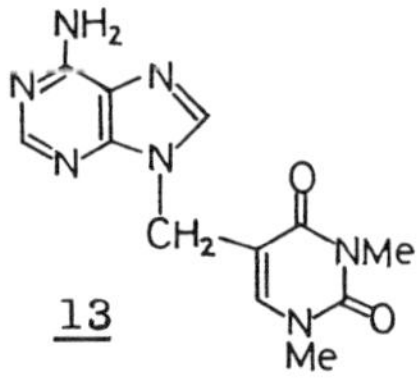

13

When 1,3-dimethylthymine (5) (5 mmol) was reacted with $Na_2S_2O_8$ (5 mmol) in the presence of adenine (2 mmol) in water (100 ml) at 80 °C for 4 h under N_2, coupling products between (5) and adenine such as (12) (0.09 mmol) and (13) (0.10 mmol) was obtained. The other products were 5-formyl- and 5-hydroxymethyl-1,3-dimethyluracils[1] (0.20 mmol) and (0.47 mmol), 5,5'-[oxybis(methylene)] bis(1,3-dimethyluracil)[1] (0.05 mmol), and recovered (5) (3.62 mmol).

REFERENCES

1. Itahara, T., Fujii, Y. and Tada, M. (1988) J. Org. Chem. 53, 3421-3424.

Electrostatic interaction in the reaction between porphyrin and DNA

Naoki Sugimoto*, Keiko Hasegawa, Noriko Monden, and Muneo Sasaki

Department of Chemistry, Faculty of Science, Konan University, Kobe 658, Japan

ABSTRACT

The association of tetrakis(4-N-methylpyridyl)porphine with salmon testes DNA in solution has been investigated in the presence of ethanol at 25 °C. It was found that the association constant was 1.14×10^{6} M^{-1} and it increased with increasing the mole fraction of added ethanol. The behavior was very different from that in the case of usual intercalaters.

INTRODUCTION

Specific interactions play very important roles on biochemical processes, but the nature of the interactions is not fully understood, for example, on the binding of drugs to nucleic acids. The technique of solvent perturbation should be useful to understand the interactions, because solvation of a drug and structure of a macromolecule are very sensitive to the environment. In this paper, we have studied spectrophotometrically the effect of ethanol on the reaction of tetrakis(4-N-methylpyridyl)porphine (TMPyP) binding to salmon testes DNA in 100 mM NaCl, 10 mM Na_2HPO_4, 0.1 mM Na_2EDTA, pH 7.0 at 25 °C.

RESULTS AND DISCUSSION

Free TMPyP had the Soret maximum at 424 nm, and the reaction with DNA led to a substantial red shift of the maximum and a large hypochromicity. The spectral shapes in ethanol-aqueous buffers were very similar to that in the aqueous buffer. The absorbance of TMPyP at 424 nm decreased with increasing DNA concentration and the minimal absorbance at the wavelength was reached by 1×10^{-4} M DNA in the aqueous buffer. The results of the titrations in the aqueous and ethanol-aqueous buffers were converted to binding isotherms. The data were analyzed with the theory (1) for site-exclusion binding using eq 1:

$$r/b = K(1-nr)^n[1-(n-1)r]^{1-n} \quad (1)$$

Here r is the average fraction of lattice binding sites of DNA occupied by TMPyP, b is the molar concentration of free TMPyP, K is the association constant of TMPyP to DNA, and n is the number of lattice residues made inaccessible by the binding of a sigle TMPyP molecule.

The obtained values of K were plotted against the mole fraction of ethanol in Fig. 1. The value of K increased with increasing the mole fraction of added ethanol. The behavior was very different from those in the case of the other drugs (2-4) binding to DNA which were intercalated between base pairs of DNA mainly due to stacking interaction. Therefore, the association of TMPyP in ethanol-aqueous solution was considered to be an external binding mainly due to electrostatic interaction. The theoretical treatment by using the polyelectrolyte model (5) supported this consideration.

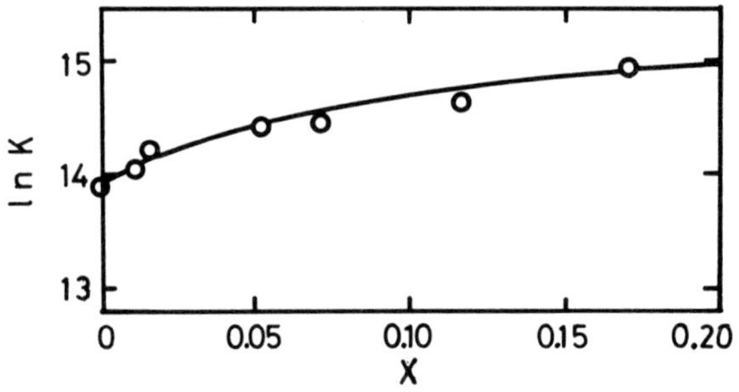

Fig. 1. The association constant, K, of the TMPyP-DNA complex vs. the molar fraction, X, of ethanol at 25 °C.

REFERENCES

1. McGhee, J. D. and von Hippel, P. H. (1974) J. Mol. Biol. **86**, 469-489.
2. Sugimoto, N., Hasegawa, K., and Sasaki, M. (1989) Chem. Express. **4**, 29-32.
3. Sugimoto, N., Monden, N., and Sasaki, M. (1989) Rep. Prog. Poly. Phys. Jpn. **32**, in press.
4. Valani, G., Torre, L. D., and Baldini, G. (1987) Biophys. Chem. **28**, 175-181.
5. Manning, G. S. (1969) J. Chem. Phys. **51**, 924-933.

Detection of single base mutation in β-thalassemia using a biotinylated oligonucleotide probe

Akio Yamane, Taturou Kawasoe, and Yukio Hattori[1]

Institute for Biotechnology Research of Wakunaga Pharmaceutical Co., Ltd., 1624 Shimokotachi, Koda-Chou, Takata-Gun, Hiroshima 729-64 and [1]Department of Clinical Laboratory Science, Yamaguchi University School of Medicine, 1144 Kogushi, Ube, Yamaguchi 755, Japan

For the detection of single base mutation in human genomic DNA, a ^{32}P-labeled allele-specific oligonucleotide (ASO) has been used successfully. To gain clinical acceptance, ASO method has been improved with the aid of PCR procedure (1) in which amplified DNA is analyzed with nonradioactive labeled ASO probe. Although the hybridization conditions of ^{32}P-labeled ASO probe were well-studied, the information about nonradioactive labeled probe is not enough for routine use. In this paper, we describe about the biotin labeled oligonucleotide probe for the detection of single base mutation.

In order to detect single base mutation in β-thalassemia, biotinylated 19-mer was synthesized and tested preliminarily. Contrary to the expectation, it was difficult to discriminate a single base mutation using biotinylated 19-mer, while the clear discrimination was obtained using ^{32}P-labeled 19-mer. The comparative experiment for the probe length (13 base, 15 base and 19 base) resulted that the 15 base was most suitable for discriminating single base mutation. For the PCR amplified sample, which is short fragment, the length of 19 base might not be necessary to recognize a unique sequence.

From the above result, biotinylated 15 base oligonucleotide probes to detect 5 alleles derived from Japanese families (2) were synthesized (Table 1). Dot hybridization was carried out for each normal and mutated probe using plasmid DNA containing normal β-globin gene as a target. Concerning the discrimination, those sets of probes were fallen into three groups. First group (Bβ11-Bβ12 and Bβ20-Bβ21) was most effective for the discrimination and

Table 1 Sequence of Biotinylated Oligonucleotide Probe

Name	Mutated Situation	Sequence
Bβ11	TATA box(-31A→G)	Bio-CTGGGCATAAAAGTC
Bβ12		Bio-CTGGGC<u>G</u>TAAAAGTC
Bβ17	codon 24(GGT→GGA)	Bio-AAGTTG<u>G</u>TGGTGAGG
Bβ18		Bio-AAGTTGG<u>A</u>GGTGAGG
Bβ17'	codon 26(GAG→AAG)	Bio-TTGGTGG<u>T</u>GAGGCCC
Bβ19		Bio-TTGGTGGT<u>A</u>AGGCCC
Bβ20	IVS II 654(C→T)	Bio-GCTATTGC<u>C</u>TTAACC*
Bβ21		Bio-GCTATT<u>A</u>CCTTAACC*
Bβ22	codon 110(CTG→CCG)	Bio-ACAGAC<u>C</u>AGCACGTT*
Bβ23		Bio-ACAGACC<u>G</u>GCACGTT*

The β-globin mutations are underlined and * shows antisense strand.

second group (Bβ17'-Bβ19) was also effective but needed prolonged washing. For the last one (Bβ17-Bβ18 and Bβ22-Bβ23), it was hard to discriminate single base mutation with any hybridization conditions. To overcome this, we introduced the competitor oligonucleotide as used in the DNA-RNA hybridization (3) and all mutations could be discriminated satisfactorily. Although the consideration about the position, length and base composition of probe sequence may help the discrimination, neighboring base sequences around mutation cannot be altered. It is consequent that the competitor oligonucleotide is very effective where the mutation is surrounded by GC rich sequence as indicated above (Bβ17-Bβ18 and Bβ22-Bβ23).

On the basis of the above results, these probes were applied to PCR amplified DNAs from clinical samples and the result thus obtained will be described.

References

1. Saiki,R.K.,Chang,C.-A.,Levenson,C.H.,Warren,T.C.,Boehm,C.D. Kazazian,Jr.H.H. and Erlich,H.A. (1988) New Engl.J.Med. 319, 537-541.
2. Hattori,Y. et.al. in preparation.
3. Nozari,G.,Rahbar,S. and Wallace,R.B. (1986) Gene 43,23-28.

dNTP imbalance and DNA double strand breaks in mouse FM3A cells and the mechanism of cell death

Yusuke Wataya, Kazuyo Watanabe, Sei Yoshida, and Akiko Hiramoto-Yoshioka

Faculty of Pharmaceutical Sciences, Okayama University, Tsushima, Okayama 700, Japan

ABSTRACT

The mechanism of intracellular deoxyribonucleoside-triphosphate (dNTP) imbalance death of mouse mammary tumor FM3A cells was studied. When the cells were exposed to 5-fluorodeoxyuridine, deoxyadenosine, or 2-chlorodeoxy-adenosine, dNTP pool imbalance resulted. The imbalance was followed by DNA double-strand breaks and subsequent cell death. The DNA double strand breaks were directly examined by means of orthogonal-field-alternation gel electrophoresis (OFAGE). Fragmented DNA band appeared to be approximately 100-200 kbp in size. The bases of 5'-termini in the DNA were cytosine and thymine. The imbalance induced endonuclase has been isolated by DEAE-agarose column chromatography.

The mechanism of intracellular deoxyribonucleoside-triphosphate (dNTP) pool imbalance induced cell death in mouse FM3A cells was investigated (1-3). We observed that 5-fluorodeoxyuridine (FdUR), deoxyadenosine and 2-chlorodeoxyadenosine induced dNTP imbalance and subsequent double-strand breaks in mature DNA accompanied by cell death. For example, the dNTP imbalance was maximal at 8 hr after 1 μM FdUR treatment; a depletion of dTTP and dGTP pools and an increase in dATP pool were observed. The imbalance was followed by breakage of mature DNA. DNA double-strand breaks as measured by the neutral elution methods were observed in the FdUR treated cells 10 hr after the administration. dNTP imbalance induced double-strand breaks were directly examined by means of orthogonal-field-alternation gel electrophoresis (OFAGE) (4). Asynchronously growing cells were treated with 1 μM FdUR for the indicated periods. In untreated cells, DNA was retained at the origin of the gel. After the administration of FdUR for 12 hr, fragmented DNA band was observed by the analysis. The band appeared to be approximately 100-200 kbp in size, which is close to the postulated size of replication units in mammalian cells (5).

For analysis of fragmented DNA, OFAGE electrophoresis of DNA in FdUR-treated cells was performed in 1.5 % low melt agarose gels in 0.6 x TBE buffer at 13 °C for 15 hr including 50 to 100 second linear gradient pulse. The band (approximately 100-200 kbp in size) was diced and labelled the 5'-end with [^{32}P]ATP and T4 polynucleotide kinase. The labelled DNA was recovered from the low melt agarose by the procedure of Hightower *et al.* (6). The isolated DNA was treated with nuclase P_1, and [^{32}P]-labelled deoxynucleoside 5'-monophosphates were identified as either deoxycytidine 5'-monophosphate and thymidine 5'-monophosphate by co-chromatography with authentic samples using cellulose thin-layer chromatography.

The double-strand breaks induced by dNTP imbalance were not repaired; both the DNA breakage and the cell death were prevented by addition of cycloheximide, an inhibitor of protein synthesis. Endonuclase activity toward double stranded DNA has been isolated from cell lysate prepared from FdUR-treated cells using DEAE-agarose column chromatography . The DNA breaks must have occurred in specific regions of the genome and we assume that the double-strand breaks of mature DNA play an important role in the process of the dNTP imbalance death. Our results indicate that imbalance of the dNTP pools is a trigger for double-strand DNA breaks and cell death.

ACKNOWLEDGMENTS

Professor Hikoya Hayatsu of the Okayama University is gratefully acknowledged for his encouragement throughout this research. This work was supported by a grant for Scientific Research from the Ministry of Education, Science and Culture of Japan.

REFERENCES

1. Yoshioka, A., Tanaka, S., Hiraoka, O., Koyama, Y., Hirota,Y.,Ayusawa, D., Seno, T., Garrett, C., and Wataya, Y. (1987) J. Biol. Chem., 262, 8235-8241.
2. Yoshioka, A., Tanaka, S., Hiraoka, O., Koyama, Y., Hirota, Y., and Wataya, Y. (1987) Biochem. Biophys. Res. Commun., 146, 258-264.
3. Hirota, Y., Yoshioka, A., Tanaka, S., Watanabe, K., Otani, T., Minowada, J., Matsuda, A., Ueda, T., and Wataya, Y. (1989) Cancer Res., 49, 915-919.
4. Carle, G. F. and Olson, M. V. (1984) Nucleic Acids Res., 12, 5647-5664.
5. Hand, R. (1978) Cell, 15, 317-325.
6. Hightower, R. C., Wong, M. L., Ruiz-Perez, L., and Santi, D. V. (1987) J. Biol. Chem., 262, 14618-14624.

Interconvertible hairpin structure found in the replication origin of phage G4 single-stranded DNA

Ichiro Hirao, Yoshifumi Nishimura[1], Miki Ishida, Kimitsuna Watanabe[2], and Kin-ichiro Miura

Department of Industrial Chemistry, Faculty of Engineering, The University of Tokyo, Hongo, Tokyo 113, [1]Graduate School of Science, Yokohama City University, Kanazawa-ku Yokohama 236, and [2]Department of Life Chemistry, Tokyo Institute of Technology, Nagatsuta, Yokohama 227, Japan

ABSTRACT
We found a synthetic GCGAAAGC fragment with a mobility greater than that of other oligodeoxyribonucleotides in gel electrophoresis to take on a stable hairpin structure possessing two terminal G-C base pairs. The GCGAAAGC sequence is also found in the replication origin of phage G4 single-stranded DNA, but the hairpin structure originally proposed differs from that of the GCGAAAGC fragment we have studied. Possibility of rearrangement of the secondary structure in the replication origin of phage G4 was examined in relation to its replication initiation mechanism.

We previously reported that short single-stranded DNA (ssDNA) fragments containing a GCGAAAGC sequence have mobilities greater than other oligonucleotides on 20% polyacrylamide-7M urea gel electrophoresis (1). This is because the GCGAAAGC fragment forms an extraordinarily stable hairpin structure comprised of GCG and AAAGC helical subsegments possibly containing two terminal G-C pairs (2)(Fig. 1a, hairpin A).

The GCGAAAGC sequence is also found in a replication origin of phage G4 ssDNA as a hairpin structure (3,4)(Fig. 1b), to which dnaG protein (primase) binds for the synthesis of its complementary primer RNA (5,6). However, the hairpin structure so far proposed (Fig. 1c, hairpin B) differs from that of the GCGAAAGC sequence we have studied (1,2). Thus, an attempt was made to determine whether the secondary structure of this hairpin region is interconvertible.

Several oligonucleotides related to the template hairpin (Fig. 1b) of the phage G4 and also hairpins A, B and C were synthesized chemically by a DNA synthesizer (Applied Biosystems model 381A). The manner of base-pairing in the hairpin structures of these oligonucleotides was clarified by T4 DNA polymerase treatment in the absence of substrate dNTPs and the

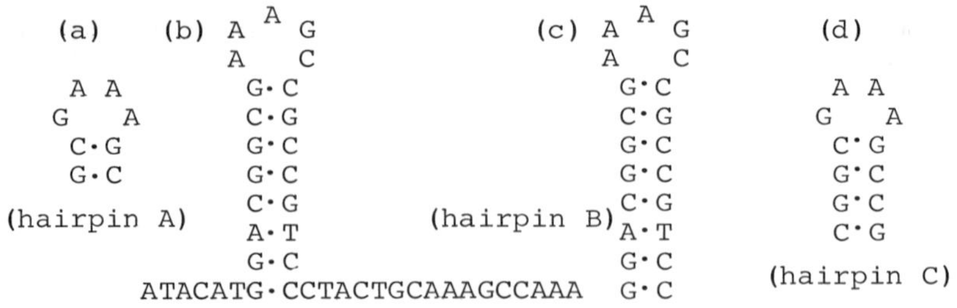

Fig.1 Hairpin structure of a synthetic fragment GCGAAAGC (a), in the replication origin of phage G4 (b) and its fragments in alternative secondary structures (c and d).

stability of the hairpin structures was assessed on the basis of melting temperature (Tm).

Neither hairpin B nor its interconvertible structure involving hairpin A (named hairpin C) was digested with T4 DNA polymerase, suggesting both fragments to form the fully base-paired stem structure shown in Fig. 1(c) and (d), since the enzyme, possessing 3'-5' exonuclease activity toward ssDNA, scarcely digests double-stranded DNA. When two terminal G-C pairs were eliminated from the hairpin B, the fragment was digested with T4 polymerase at the 3'-terminal, indicating the fragment to no longer possess a fully base paired stem structure, but has an extra single-stranded tail at the 3'-terminal. This indicates that a structural rearrangement quite likely occurs to give rise to an extraordinarily stable core structure of hairpin A. In fact, the Tm of hairpin B was 3°C lower than that of hairpin C, even though hairpin B has four more base pairs than hairpin C.

Based on the interconvertible hairpin structure above, initiation mechanism of DNA replication will be discussed.

REFERENCES

1. Hirao, I., Naraoka, T., Kanamori, S., Nakamura, M. and Miura, K. (1988) Biochem. Int. 16, 157-162.
2. Hirao, I., Nishimura, Y., Naraoka, T., Watanabe, K., Arata, Y. and Miura, K. (1989) Nucleic Acids Res. 17, 2223-2231.
3. Godson, G. N., Barrell, B. G., Staden, R. and Fiddes, J. C. (1978) Nature 276, 236-247.
4. Sims, J., Capon, D. and Dressler, D. (1979) J. Biol. Chem. 254, 12615-12628.
5. Rower, L. and Kornberg, A. (1978) J. Biol. Chem. 253, 770-774.
6. Bouche, J.-P., Zechel, K. and Kornberg, A. (1975) J. Biol. Chem. 250, 5995-6001.

Study on the function of the bent DNA within adenovirus type 2 EIA enhancer: analysis using an *in vitro* transcription system

Takashi Ohyama, Masahiko Nagumo, and Sadatoshi Sakuma

Section of Molecular Oncology, Meiji Institute of Health Science, 540 Naruda, Odawara 250, Japan

ABSTRACT

The role of the bent DNA within human adenovirus type 2 (Ad2) enhancer in transcriptional activation of E1A gene has been investigated. The bent region of the enhancer was substituted by a straight DNA fragment or by a bent DNA fragment of 194 bp from origin region of phage λ replication. Employing an *in vitro* transcription system using HeLa cell extract, effects of mutant enhancers on E1A gene transcription were analyzed. The results obtained showed that the amount of transcripts is affected by changes in local conformation in and around the enhancer.

DNAs having curved trajectory of their helix axes are called bent DNA or curved DNA, and have been found in DNA molecules from various sources, mostly in the origin region of replication in both prokaryotic and eukaryotic organisms. Such structures have also been found in the transcriptional control regions of some genes (1～4). The functional significance of these structures, however, remains still obscure. We have found a bent DNA within Ad2 enhancer and determined the location of the structure (4). The bent DNA was revealed to exist in the upstream half of the putative enhancer/packaging region of Ad2 DNA, where, from the sequence homology to Ad5 (5,6), multiple enhancer elements for transcription and the elements required for packaging of Ad2 DNA are thought to be contained. To determine whether this unusual DNA structure is involved in the transcriptional activation process of EIA gene, *in vitro* transcription assays were carried out.

The plasmid pP5 was firstly constructed to run the *in vitro* system. This was prepared by cloning left terminal *Ava*I fragment of Ad2 DNA (757 bp) into the *Eco*RI and *Ava*I sites of plasmid pUC19, and *Taq*I fragment (629 bp) into the *Pst*I and *Acc*I sites, respectively (The *Taq*I fragment has a *Pst*I site at the other end and a 10 bp fragment between this site and the left end of Ad2 DNA.

The *Eco*RI site of the *Ava*I fragment was created after destroying the *Pst*I site; see ref. 4). Transcription of EIA gene starts at nucleotide 498, so when the plasmid is digested with *Bam*HI (*Bam*HI site exists between *Ava*I site and *Acc*I site of pUC19 and this is a unique site in pP5) and the transcription by HeLa cell extract is performed, then transcripts of 269 and 147 nucleotides, derived from transcriptions by E1A promotors in the *Ava*I and *Taq*I fragments respectively, are expected to be produced. When transcription was carried out with [α-^{32}P] UTP, expected transcripts were proven to accumulate.

In the next step, plasmids carrying both a wild type and a mutant enhancer were constructed by modifying the structure of pP5. That is, the Ad2 *Ava*I fragment in pP5 was altered to have a mutant enhancer by substituting the bent region within it by a straight DNA fragment or by a bent DNA fragment of 194 bp from origin region of phage λ replication. Then, each plasmid was digested with *Bam*HI and transcription assays were carried out with the same procedure as in the case of pP5. Comparing the amounts of the transcripts produced (269 nucleotides, from mutant enhancers; 147 nucleotides, from the wild type enhancer), effects of mutant enhancers on the transcription of EIA gene were analyzed.

It was revealed that substitutions of the bent region resulted in about 40-70% decrease in levels of truncated EIA transcripts. In decreasing transcription levels, substituted bent DNA from phage λ showed obviously different effects depending on its orientation, indicating that the local conformation in and around the enhancer is likely to be involved in the transcription process of EIA gene.

REFERENCES

1. Bossi, L. and Smith, D.M. (1984) Cell 39, 643-652.
2. Mizuno, T. (1987) Gene 54, 57-64.
3. Inokuchi, K., Nakayama, A. and Hishinuma, F. (1988) Nucl. Acids Res. 16, 6693-6711.
4. Ohyama, T. and Hashimoto, S. (1989) Nucl. Acids Res., in press.
5. Hearing, P. and Shenk, T. (1986) Cell 45, 229-236.
6. Hearing, P., Samulski, R.J., Wishart, W.L. and Shenk, T. (1987) J. Virol. 61, 2555-2558.

Signal structure of the *cis*-acting element recognized by VirG protein, a positive regulator in *Agrobacterium*

Takashi Aoyama, Mituru Takanami, and Atsuhiro Oka

Institute for Chemical Research, Kyoto University, Uji, Kyoto 611, Japan

ABSTRACT

VirG protein is known to be the positive regulator for vir genes of Ri and Ti plasmids in Agrobacterium. To investigate the cis-acting element recognized by VirG, we have determined the transcriptional start points of the vir genes on Ri plasmid (pRiA4b). From the analysis of the upstream sequences, it has been found that the sequences generally similar to 5'TG(A/T)AA(C/T)3' appear in phase with an 11-base pair interval, and that the -35 and -10 regions of the promoters are located nearly in the reversal phase with those phasing sequences. These sequence characteristics are conserved in the vir genes on Ti plasmids (pTiA6 and pTiC58). Moreover, the similar sequence structures were found in the regulatory regions of the genes controlled by PhoB and OmpR, which are the E. coli positive regulators closely related to VirG in their amino-acid sequences.

Agrobacterium harboring pTi or pRi confers tumor on plants by transferring a specific segment (T-DNA) of the plasmid to plant nuclear genomes. The genes required for the T-DNA transfer, vir genes, are inducibly transcribed in response to plant phenolic compounds such as acetosyringone. In this response, a two-component regulatory system composed of VirA and VirG proteins is thought to be concerned with the signal transduction. VirG shares the similarity of amino-acid sequence over its entire length with PhoB and OmpR, positive regulators of two-component regulatory systems in E. coli. Like PhoB and OmpR, VirG is supposed to bind to specific sequences in the upstream region for transcriptional activation. However, the cis-acting element recognized by VirG has not been determined so far. For vir genes on Ti plasmid (pTiA6 and pTiC58), comparison of the upstream sequences suggested a 12-base pair (bp) sequence (5'TNCAATTGAAAPy3') to be the consensus sequence which should be recognized by VirG. Although sequences similar to the 12-bp sequence commonly exist in the upstream region of the vir genes, their positions relative to the transcriptional start points are different with the vir genes, and their validity has not been made evident.

In this report, we determined the transcriptional start points of vir gene on Ri plasmid (pRiA4b) by using the methods of the S1 mapping and the primer extension. The transcripts examined were those from virA, virB, virC, virD and virG. In order to find the cis-acting element, we searched the upstream regions of the inducible transcripts for commonly existing sequence by a computer. As the result, a 6-bp sequence of 5'TG(A/T)AA-(C/T)3' was found in all of the upstream regions. It was also found that the sequences generally similar to the 6-bp sequence appear at a high frequency in the regions and that almost all of them are located in phase with an 11-bp interval. Furthermore, the -35 and -10 regions of all the promoters were located nearly in the reversal phase with these phasing sequences, while the distance between the promoters and the phasing sequences varied with the vir genes. These structural characteristics are also conserved in Ti plasmids with a few exceptions, suggesting the biological significance for the transcriptional activation. From these circumstances, we propose a model that if a VirG molecule once interact at one of the phasing sequences, additional VirG proteins can bind cooperatively in head-to-tail manner along DNA by an 11-bp phase regardless of the presence or absence of the phasing sequence, thereby the promoter region become competent for RNA polymerase binding.

Since PhoB and OmpR are closely related to VirG in their amino-acid sequences, it is expected that the sequence structure found in vir genes shares some common feature with those recognized by PhoB and OmpR. Actually, 6-bp phasing sequences of 5'T(G/T)TCA(C/T)3' and 5'TGAAAC3' have been found in the upstream regions of the genes regulated by PhoB and OmpR, respectively. The relative phases of the -35 and -10 regions of the promoter to the phasing sequences are also nearly reversal as in the case of vir genes.

A polymorphic DNA marker genetically linked to congenital malformations

Tetsuya Moriuchi, Shoko Awataguchi, Yasushi Taniguchi, Katsuhiko Iwasaki[1], Motoo Katabami[2], Hiroyuki Naito[2], and Noboru Kuzumaki[2]

Department of Cell Biology and [1]Department of Gynecology, Tokai University School of Medicine, Isehara 259-11 and [2]Laboratory of Molecular Genetics, Cancer Institute, Hokkaido University School of Medicine, Sapporo 060, Japan

ABSTRACT

Unusual restriction fragments were detected by DNA blot hybridization with PCNA (DNA polymerase-delta auxiliary protein) probe in one of seven cases of congenital malformations. Chromosomal in situ hybridization localized PCNA gene to region q31-35 of human chromosome 2. To discover the locus more closely associated with congenital malformations, a cloned DNA segment which has been mapped to chromosomal region 2q33-36 was tested for restriction fragment length polymorphisms (RFLPs) in these patients. The 2q33-36 probe hybridized with 2.1-kb, 1.9-kb and 1.7-kb fragments in ten normal control samples. In seven cases of congenital malformations examined, however, the band of 2.1 kb is absent in six cases and the band of 1.7 kb in one case. These results indicate that the locus closely linked to congenital malformations is present in the proximity of PCNA locus.

INTRODUCTION

Congenital malformations are structural defects present at birth. They occur in both hereditary and nonhereditary forms. In spite of numerous chromosomal studies the common defect in congenital malformations has never been detected. At present there is no DNA marker which could lead to a screening test for gene carriers of the congenital malformation. We have now identified a polymorphic DNA marker that shows close genetic linkage to the congenital malformation gene.

MATERIALS AND METHODS

The patients studied were aged between 0 and 9 years old and had a variety of malformations such as anencephaly, microcephaly, spina bifida, hemivertebra, atresia ani, stenosis of aortic valve, and cleft lip and palate. DNA was isolated from placentas,

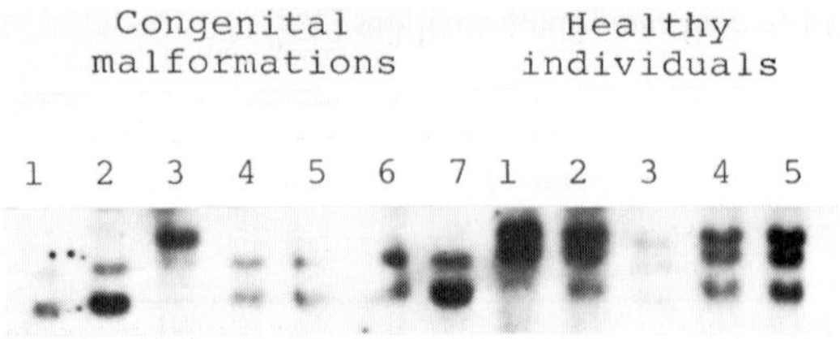

Fig. 1

Hybridization of the 2q33-36 probe to restriction enzyme-digested human genomic DNA.

The 2q33-36 probe reveals a RFLP with fragments of 2.1 kb, 1.9 kb and 1.7 kb.

post-mortem tissues or peripheral leukocytes by the method of SDS-proteinase K and phenol/chloroform treatment. DNA probes used are rat PCNA (DNA polymerase-delta auxiliary protein) cDNA (1, 2) and a cloned DNA segment which has been mapped to 2q33-36 (2q33-36 DNA fragment).

RESULTS AND DISCUSSION

We first analysed DNA polymorphisms in a variety of genetical disorders by blot hybridization with PCNA probe. Unusual restriction fragments were detected in one of seven cases of congenital malformations. Chromosomal *in situ* hybridization localized PCNA gene to chromosomal region 2q31-35, with a higher possibility for localization at 2q33 (3). On the assumption that the locus more closely associated with congenital malformations may be present in the proximity of PCNA locus, patient DNA samples were screened for polymorphisms with a DNA fragment which has been mapped to chromosomal region 2q33-36. The 2q33-36 probe hybridized to 2.1-kb, 1.9-kb and 1.7-kb fragment in ten normal control samples, while the upper band (2.1 kb) was absent in six of seven cases of congenital malformations and the lower band (1.7 kb) in one case (Fig. 1). These results indicate that 'congenital malformation locus' may be in close proximity to PCNA locus and that this DNA fragment will be useful in the study of morphogenesis during embryogenesis.

REFERENCES

1. Matsumoto, K. et al. (1987) EMBO J. 6, 637-642.
2. Prelich, G. et al. (1987) Nature 326, 517-520.
3. Yoshida, M. C. et al. (1989) Jpn. J. Human Genet. 34, 32.

Sequence specificity of the human autonomously replicating sequence binding protein

Shunji Izuta, Motoshi Suzuki, Kiyohide Kojima, and Shonen Yoshida

Research Institute for Disease Mechanism and Control, Nagoya University School of Medicine, Showa-ku, Nagoya 466, Japan

ABSTRACT

We have previously found that the protein which bound to the essential region of human autonomously replicating sequences (ARS) could be the c-*myc* proto-oncogene product. Here we examined the binding specificity of human ARS binding protein purified from Burkitt's lymphoma Raji cells. Among several oligodeoxynucleotides, this protein bound to the fragments containing the sequence of 5'-CAPyCTCTNA-3'. Competition analysis revealed that the ARS binding protein could recognize not only the nucleotide sequences but also the high ordered structure.

INTRODUCTION

Autonomously replicating sequences (ARS) are thought to act as the origin of chromosomal DNA replication. Human ARS was found at the 5'-upstream of c-*myc* gene[1], and we have recently found that the protein which bound to the essential region (termed mycSL1) of human ARS could be the c-*myc* protein[2]. For clarifying the mechanism of replication initiation, it is important to study this protein in detail. From this context, we had examined its binding specificity.

RESULTS AND DISCUSSION

The binding ability of human ARS binding protein to several oligodeoxynucleotides was examined using gel retardation assay. This protein bound to the mycSL1 (essential region of ARS) and mycSL2 (another region of ARS), but failed to bind to $(dA)_{10}$-$(dT)_{10}$ or pUC118 DNA fragments (Fig. 1a). This protein also tightly bound to the one of

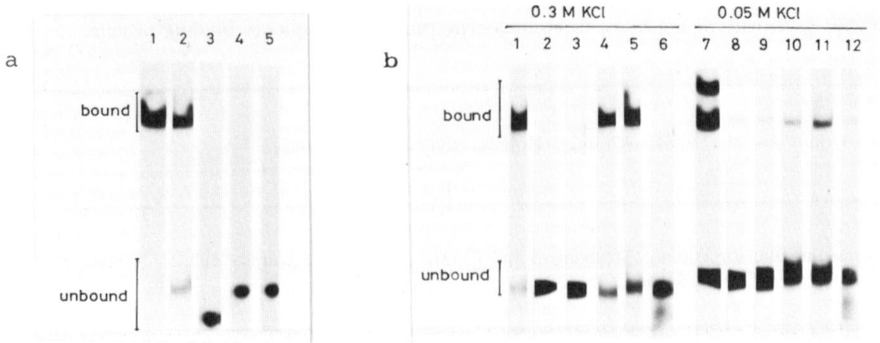

Fig. 1. (a) Gel retardation analysis of the binding ability of human ARS binding protein to several 5'-^{32}P oligodeoxyribonucleotides (1 ng). Lane 1; mycSL1 (36bp), 2; mycSL2 (36), 3; $(dA)_{10}$-$(dT)_{10}$, 4; pUC118 EcoRI-BamHI (21), 5; pUC118 BamHI-HindIII (30). (b) Competition of the formation of protein-^{32}P mycSL1 (1 ng) complex by several DNAs (100 ng). Lane 1,7; without competitor, 2,8; mycSL1, 3,9; mycSL2, 4,10; pUC118, 5,11; pBR322, 6,12; HaeIII digested pUC118 fragments.

HindIII-KpnI digested SV40 DNA fragments (752bp). These bound fragments contained the common sequence of 5'-CAPyCTCTNA-3' (N=T,G), therefore this protein could dominantly recognize this nonamer. Fig. 1b shows the result of competition analysis. The mycSL1 and 2 competed with the mycSL1 for the ARS binding protein, but pUC118 and pBR322 did not in the presence of 0.3 M KCl. However, in the low salt condition, both plasmids competed with mycSL1. Also the HaeIII digested pUC118 fragments competed with mycSL1 either in the presence of 0.3 M KCl or in the low salt condition. The structure of plasmid is known to change according to the salt concentration, whereas that of linearized molecule is not affected by the salt. These results indicate that the human ARS binding protein could recognize not only the sequences but also the high ordered structure.

REFERENCES

1. Iguchi-Ariga, S.M.M., Ozaki, T., Itani, T., Ogata, M., Sato, Y. and Ariga, H. (1988) EMBO J., 7, 3135-3142.
2. Izuta, S., Suzuki, M., Kojima, K. and Yoshida, S. submitted.

Stability of CAT gene transcript depends on the 3′-end structure

K.Sato*, M.Yagi, T.Nohno[1], K.Kobayashi, and K.Hosokawa

Department of Biochemistry and [1]Department of Pharmacology, Kawasaki Medical School, Kurashiki 701-01, Japan

ABSTRACT

To determine the involvement of 3'-end structure of mRNA in the regulation of gene expression in eukaryotic cell, a series of pSCAT plasmids was constructed with chloramphenicol acetyltransferase (CAT) gene and the 3'-reguratory elements from various eukaryotic genes. As a result of determination of the CAT activities and the mRNA level generated from these plasmids, the both results were well correlated. Forethermore, the treatment of transfected cells with phorbol ester (TPA) revealed that the polyadenylated CAT mRNAs form some genes were stabilized, in contrast, the mRNAs bearing 3'-end structure of histone or C-myc genes were not.

INTRODUCTION

The regulation of eukaryotic mRNA turnover can play a major role in controlling gene expression and in determining cellular mRNA levels (1). One of important functions of the poly(A) tract may be to influenced mRNA stability. Among the poly(A) mRNAs, one class of transcripts including a β-globin mRNA is very stable, and has a half life of greater than 17 h (2). While the other genes, such as numerous lymphokine, cytokine or proto-oncogenes generate short-lived mRNAs which have half lives of 30 min or less (3, 4, 5). Well-conserved AU-rich sequences were found in the 3'-untranslated region of mRNA from the transiently expressed genes, and involved in the regulation of mRNA degradation (6). On the other hand, the histone gene expression are regulated during cell cycle (7), and the turnover rate of histone mRNA is relatively rapid.

In this paper, we determined whether the stability of CAT gene transcripts was controlled by placing the 3'-untranslated region from various genes. Plasmids, pSCAT6 and pSCAT10 used in this study contain a multipl insertion sites at the downstream of the CAT gene (8).

RESULTS AND DISCUSSION

CAT Gene Expression Influenced by 3'-End Structure

KB cell was transfected with a series of pSCAT plasmids, and the CAT activity was determined. The plasmids containing poly(A) signal or histone 3'-region in correct orientation generated high CAT activities. On the other hand, low CAT activities were exhibited by the plasmids which were lack of the poly(A) region or inserted that in anti-sense direction.

mRNA Stability in Transfected Cell

Transcripts bearing SV40 poly(A) tail (from pSV2CAT and pSCAT10) were stable more than pSCAT6-H2 mRNA carrying the 3'-region from histone gene. Plasmid pSCAT6 produced extremely short-lived transcript, due to lack of poly(A) tail. Compared with these transcripts, mRNAs from plasmids bearing LTR of RSV or poly(A) region of gastrin gene were very stable.

Effects of TPA Treatment on mRNA Stability

TPA-treatment of cell increase stability of some genes, for example, a human granulocyte-monocyte colony stimulating factor (GM-CSF) gene or a human c-fos oncogene (5, 6). The transcripts containing poly(A) track derived from SV40 early gene (pSV2CAT) or adenovirus 5 E1B gene (pSCAT6-E1B4) were relatively short-lived, but extremely stabilized by TPA stimulation. On the other hand, mRNAs transcribed from pSCAT6-H2, pSCAT6-cmyc12 and histone gene made no responce to TPA-treatment. Furthermore, pSCAT6-RL24 (RSV-LTR), pSCAT6-G17 (gastrin 3'-region) and β-actin gene produced extremely stable transcripts which turnover rate was not depend on the presence of TPA. These results suggested a possible mechanism of mRNA turnover involving of degradation factor(s).

ACKNOWLEDGMENTS
This work was supported by the grant from Kawasaki Medical School.

*Present address: Division of Biochemistry, Institute of Steroid Research, Tottori University School of Medicine, Yonago 683, Japan

REFERENCES
1. Ross,J. (1988) Mol. Biol. Med. 5, 1-14
2. Aviv,H. et al. (1976) Cell 8, 496-503
3. Raj,N.B.K. and Pitha,P.M. (1981) Proc. Natl. Acad. Sci. 78, 7426-7430
4. Greenberg,M.E. and Ziff,E.B. (1984) Nature 311, 433-438
5. Sariban,E. et al. (1988) Mol. Cell. Biol. 8, 340-346
6. Shaw,G. and Kamen,R. (1986) Cell 46, 659-667
7. Sittman,D.B. et al. (1983) Proc. Natl. Acad. Sci. USA. 80, 1849-1853
8. Sato,K. et al. (1986) Mol. Cell. Biol. 6, 1032-1043

Phosphoseryl–tRNA in *Escherichia coli*

Takaharu Mizutani, Naosuke Maruyama, and Hiroshi Kurata

Faculty of Pharmaceutical Sciences, Nagoya City University, Mizuho-ku, Nagoya 467, Japan

ABSTRACT

Two seryl-tRNAs, which were prepared from natural suppressor tRNA-rich fractions on a pattern of chromatography on Sephadex A-50, were phosphorylated by a tRNA kinase in *Escherichia coli* B. A part of phosphate on the tRNA was confirmed to be phosphoserine. Phosphoseryl-tRNA is universal in bacteria and vertebrates. Phosphoseryl-tRNA should be an intermediate from seryl-tRNA to selenocysteyl-tRNA.

INTRODUCTION

In *Escherichia coli*, it was believed that there was no natural suppressor tRNA, however it was suggested that *selC*, relating to the synthesis of formate dehydrogenase of a selenoprotein, coded for the tRNA, even though the presence of the tRNA is not clarified[1]. We detected the product of *selC* in the tRNA preparation of *E. coli* MC 4100 and wild type strains by the dot blot hybridization method with a DNA probe (ACCGCTGGCGGC) corresponding to the extra arm of *selC* tRNA[2]. This probe hybridized with the tRNA preparation of *E. coli* but not with bovine and yeast tRNA. The content of the tRNA hybridizing with the probe was estimated to be 0.02% in total tRNA. It was found that two tRNA peaks hybridizing with the probe on the chromatographic pattern of tRNA on Sephadex A50[2].

RESULTS AND DISCUSSION

Two fractions hybridizing with the probe were both phosphorylated with a tRNA kinase (ATP:seryl-tRNA phosphotransferase) in *E. coli* B. No phosphorylation was found on the non-acylated suppressor tRNA. The tRNA fraction not hybridizing with the probe was not phosphorylated. This showed that serine residues on suppressor seryl-tRNA was phosphorylated to become phosphoserine. It was confirmed by analyses of phosphorylated-tRNA that a part of phosphate was to be phosphoserine. The total tRNA preparation of E. coli W was phosphorylated and chromatographed on Sephadex A50. The pattern showed two peaks of phosphate, coincident with the results of hybridization. Phosphate on this phosphorylated-tRNA was also confirmed as phosphoserine by analyses on paper electrophoresis. From the ratio of [^{3}H]-phosphoseryl-tRNA to total [^{3}H]seryl-tRNA, the level of suppressor tRNA was estimated to be about 0.02% of the total tRNA. This value agreed with that of hybridization and is near the level of mammalian suppressor tRNA. Thus, this paper clarified that, in both E. coli and vertebrates, the normal opal suppressor seryl-tRNA was converted to phosphoseryl-tRNA. This phosphoseryl-tRNA should be an intermediate from suppressor seryl-tRNA to selenocysteyl-tRNA in some species[1, 3, 4] as well as in mammals[5].

REFERENCES

1. Leinfelder, W., Zehelein, E., Mandrand-Berthelot, M. and Böck, A. (1988) Nature 331, 723-725.
2. Maruyama, N., Hitaka, T., Mizutani, T. and Sukenaga, Y.(1988) Abstract of the 11th meeting of the Japanese Molecular Biology Society, p113.
3. Menon, N.K., Peck, H.D., Gall, J.L. and Przybyla, A.E.(1988) J. Bacteriol. 170, 4429.
4. Yamazaki, S(1982)J. Biol. Chem. 257, 7926-7929.
5. Mizutani, T. and Hitaka, T.(1988)FEBS Letters 232, 243-248.

O-Alkyl *O*-nucleoside 3′-phosphonates as novel starting materials for oligonucleotide synthesis

Takeshi Wada, Ryohei Kato, Hitoshi Hotoda, Mitsuo Sekine, and Tsujiaki Hata*

Department of Life Chemistry, Tokyo Institute of Technology, 4259 Nagatsuta, Midoriku, Yokohama 227, Japan

ABSTRACT

It was found that the title compounds were activated rapidly into the corresponding highly reactive phosphorochloridites by use of tris(2,4,6-tribromophenoxy)dichlorophosphorane (BDCP) as a chlorinating reagent. The reaction was applied to the synthesis of oligodeoxyribonucleotides.

Alkyl nucleoside phosphonates are versatile intermediates in the synthesis of oligonucleotides. In the H-phosphonate method, internucleotidic phosphonate diester linkages could be transformed into various phosphate analogues (1). It is well known that certain "positive halogen" compounds react with dialkyl phosphonates to generate the corresponding phosphoryl halides with oxidation of phosphorus atom (2). Recently, we have described a new reaction, non-oxidative halogenation of dialkyl phosphonates to form dialkyl phosphorochloridites by use of tris(2,4,6-tribromophenoxy)dichlorophosphorane (BDCP) as a chlorinating reagent (3). In this study, we wish to report the application of the reaction to the synthesis of oligodeoxyribonucleotides.

Alkyl nucleoside phosphonates (**1a**; R= 2-cyanoethyl, **1b**; R= methyl, **1c**; R= 2,2,2-trichloroethyl) were prepared as follows. To a well-stirred solution of alkyl phosphorodichloridites in THF at -78° C was added dropwise a solution of a suitably protected nucleoside, 1,2,4-triazole, and pyridine in THF for 10 min. After the mixture was stirred for 30 min, successive hydrolysis and extraction gave pure **1a-c** in high yield.

The activation reaction of **1** with BDCP was monitored by ^{31}P-NMR. 2-Cyanoethyl 5'-O-dimethoxytrityl-N^{3}-benzoylthymidine

3'-phosphonate (**1a**) was mixed with 1.5 equiv of BDCP (-64.62 ppm) in pyridine at room temperature. After 5 min, **1a** (7.94 ppm) was completely converted into the corresponding phosphorochloridites (**2a**, 165.27 ppm) along with the formation of tris(2,4,6-tribromophenyl) phosphate (-23.01 ppm). Next, 30 fold excess of 2 activated 1 with BDCP in situ was allowed to react with N4-anisoylcytidine on the CPG (3) for 5 min. After usual oxidation, the dinucleoside phosphate on the CPG (5) was obtained in 99% yield. In a similer manner, fully protected dC_{10} was obtained in 73% yield. The average coupling yield was 97%. The fully protected 10 mer was deprotected and purified by the conventional procedure to give pure dC_{10} in 27% yield.

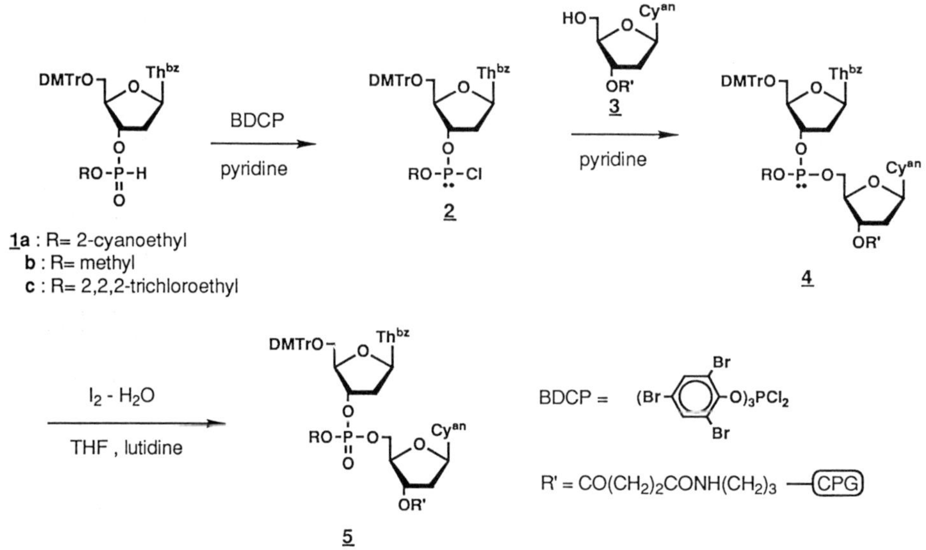

REFERENCES

1. For example; a) Kume, A., Fujii, M., Sekine, M., and Hata, T. (1984) J. Org. Chem., 49, 2139-2143. b) Froehler, B. C. (1986) Tetrahedron Lett., 27, 5575-5578. c) de Vroom, E., Spierenburg, M. L., Dreef, C. E., van der Marel, G. A., and van Boom, J. H. (1987) Recl. Trav. Chim. Pays-Bas, (1987) 106, 65-66.
2. Miller, B. (1965) In Grayson, M. and Griffith, E. J. (ed.) Topics in Phosphorus Chemistry, Wiley, New York, Vol. 2, pp 133-199
3. Wada, T., Hotoda, H., Sekine, M., and Hata, T. (1988) Tetrahedron Lett., 29, 4143-4146.

Synthesis of a linker, d(GGAATTCC), through liquid-phase and phosphoroamidate approaches involving deamidation with tetrabutylammonium nitrite−acetic anhydride

Shigeyoshi Nishino, Yasuhiro Nagato, Yoshihiro Hasegawa, Kazuo Kamaike[1], and Yoshiharu Ishido[1]

Department of Chemistry, Faculty of Science, Tokyo Institute of Technology, O-okayama, Meguro-ku, Tokyo 152 and [1]Laboratory of Pharmaceutical Chemistry, Tokyo College of Pharmacy, Horinouchi, Hachi-ohji, Tokyo 192-03, Japan

ABSTRACT

The title linker was synthesized in the manner of dimer-units introduction through liquid-phase and phosphoroamidate approaches involving deamidation with tetrabutylammonium nitrite - acetic anhydride.

INTRODUCTION

A novel approach to deamidation of phosphoroamidates with nitrites - acetic anhydride has been reported in the previous symposium,[1] in which we described an essential aspect of this approach that we should use succinyl protecting group for the amino groups on the nucleic acid base moieties in order to completely remove their active protons due to the highly reactive property of mixed anhydride of acetic and nitrous acids. This approach made it possible to remove the phosphoroamidate protecting group instantly(1-2minutes), and we could thus constructed the title octamer up to the corresponding hexamer, d[GGAATT]; whole aspect of the synthesis will be described herein.

RESULTS AND DISCUSSION

Deanisidation of a 5'-*O*-dimethoxytrityl-2'-deoxyribonucleosides 3'-phosphoroanisidates (**1**)[All the amino groups involved in the heterocyclic moieties, except thymidine, should be protected with succinyl group in order to keep them free from nitrosation reaction] by the use of tetrabutylammonium nitrite (3 mol. equiv.) - acetic anhydride (4 mol. equiv.) in pyridine was efficiently induced at room temperature, being over within 1 - 2 minutes. The resulting phosphodiesters were subsequently purified through reversed-phase column chromatography after extraction and isolated as the triethylammonium phosphodiesters (**2**). On the other hand, 5'-*O*-dimethoxytrity group was removed by the treatment of **1** with *p*-toluenesulfonic acid monohydrate in methanolic dichloromethane to give **3** according to the usual manner.

Deanisidation of di-2'-deoxyribonucleotides 3'-phosphoroanisidate, which were synthesized by the coupling

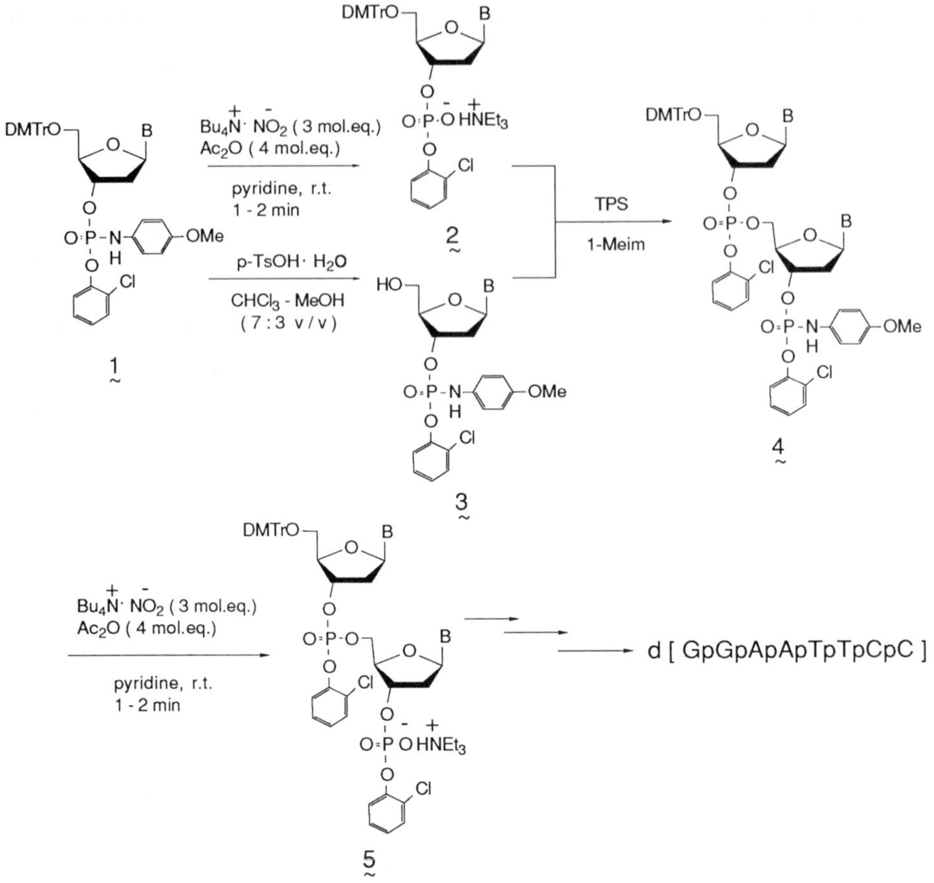

reaction of 2 with 3 by the use of 2,4,6-triisopropyl-benzenesulfonyl chloride (TPS) - 1-methylimidazole (1-Meim), with tetrabutylammonium nitrite - acetic anhydride in pyridine afforded the corresponding triethylammonium 3'-phosphodiesters (5) efficiently.

The resulting triethylammonium salts of dinucleotides 3'-phosphodiesters (5) were coupled in the stepwise manner of dimer-units introduction, i.e., d[TT], d[AA], and, then, d[GG] onto d[CC], to give the title octamer of d[GGAATTCC] successfully.

REFERENCES

1. Nishino, S., Nagato, Y., Hasegawa, Y., Kamaike, K.,and Ishido, Y. (1988) Nucleic Acids Res. Symp. Ser., 20, 73-74.

A phosphorbisamidite approach to the synthesis of oligodeoxyribonucleotides and their analogues

Kazushige Yamana, Yoshitaka Nishijima, Hidehiko Nakano, Osamu Sangen, Hiroaki Ozaki[1], and Takeo Shimidzu[1]

Department of Applied Chemistry, Himeji Institute of Technology, 2167 Shosha, Himeji 671-22 and [1]Division of Molecular Engineering, Graduate School of Engineering, Kyoto University, Sakyo-ku, Kyoto 606, Japan

ABSTRACT Utilities of deoxyribonucleoside 3'-O-phosphorbisdiethylamidites in the synthesis of oligodeoxyribonucleotides and their analogues are described.

Recent reports[1,2,3)] have shown that a deoxyribonucleoside phosphorbisamidite as a new class of intermediate is useful in the synthesis of oligodeoxyribonucleotides and their analogues possessing modified backbone. Essential feature of this approach involves that the chain elongation reactions proceed via phosphoramidite followed by H-phosphonate linkages both of which can lead to various backbone modifications of oligodeoxyribonucleotides.

We now wish to report the outline of our approach using deoxyribonucleoside phosphorbisdiethylamidites 3 to the synthesis of oligodeoxyribonucleotides and their analogues. A convenient procedure for the preparation of 3 is based on the selective phosphitylation of 5'-O-dimethoxytrityl deoxyribonucleoside with tris(diethylamino)phosphine 1 in the presence of tetrazole and diethylamine.[3)] The intermediate 3 thus obtained could be used without purification for the solid-phase synthesis of oligodeoxyribonucleotides. The activation of 3 by using

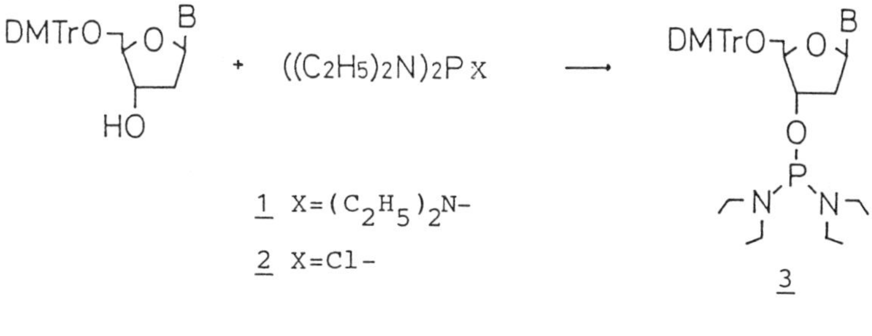

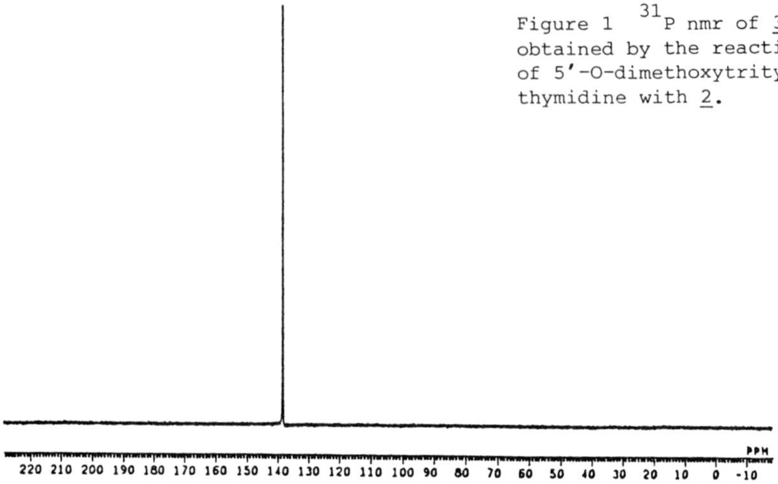

Figure 1 ^{31}P nmr of 3 obtained by the reaction of 5'-O-dimethoxytrityl thymidine with 2.

p-nitrophenyltetrazole results in the efficient coupling with 5'-hydroxyl group of nucleoside on a solid support. For example in the synthesis of dCCTAGCTAGG, each coupling efficincy was estimated to be 97 %. The oligomer was identical on hplc and tlc to the authentic sample synthesized by using β-cyanoethylamidite chemistry and completely hydrolyzed by snake venom phosphodiesterase.

The preparation of 3 by the reaction of 5'-O-dimethoxytrityl deoxyribonucleotide with bis(diethylamino)phosphorchloridite 2 is found to be more promising. Typically, 5'-O-dimethoxytrityl thymidine was allowed to react with 2(equiv) in dry CH_2Cl_2 containing 3 equivs of diethylamine. After the aqueous exstraction of the reaction mixture, the intermediate 3 was obtained in a pure form(Figure 1). In the synthesis of thymidine tetramer using 3 activated by p-nitrophenyltetrazole, each coupling efficiency exceeded 99 %. Furthermore, 3 can be applied to the automated synthesis by the slightly modified standard protocol. Preliminaly experiments show that T_{40} can be obtained in a avarage yield of over 97 %.

Another possibilities of our approach involve the synthesis of oligodeoxyribonucleotides with modified backbone. It was found that the phosphordiethylamidite formed after the coupling can be converted easily by t-butyl hydroperoxide to the corresponding amidate.

REFERENCES
1) B.Uznanski, A.Will, and W.J.Stec(1987) Tetrahedron Lett.,28, 3401.
2) J.Nielson, W.K.D.Bill, and M.H.Caruthers(1988) ibid., 29, 2911.
3) K.Yamana, Y.Nishijima, A.Oka, H.Nakano, O.Sangen, H.Ozaki, and T.Shimidzu(1989) Tetrahedron, in press.

Oligoribonucleotide synthesis by use of the 2-(methylthio)-phenylthiomethyl (MPTM) group

Mitsuo Sekine* and Takeshi Nakanishi

Department of Environmental Chemistry and Engineering, Tokyo Institute of Technology, Nagatsuta, Midoriku, Yokohama 227, Japan

Abstract

The 2-(methylthio)phenylthiomethyl (MPTM) group was developed as a new type of 2'-hydroxyl protecting group in oligoribonucleotide synthesis. The building monomer units of uridine and cytidine for the phosphotriester approach were synthesized from 2'-O-(1,3-benzodithiol-2-yl)-3',5'-O-(1,1,3,3-tetraisopropyldisiloxan-1,3-diyl)uridine and successfully utilized for the synthesis of CpUpG.

Introduction

In our continuing study on the synthesis of O-methylated deoxyribonucleosides,(1) we have recently found that the 1,3-benzodithiol-2-yl (BDT) group attached to the 3'- or 5'-oxygen of thymidine could be converted to the 2-(methylthio)phenylthiomethyl (MPTM) group by treatment with tributyltin hydride (TBTH) in the presence of 2,2'-azobis(isobutyronitrile) (AIBN) followed by S-methylation with methyl iodide.(2) The MPTM group was found to be stable under acidic and basic conditions such as 80% acetic acid and 2 M NaOH but removable upon treatment with mercury (II) chloride in aqueous acetonitrile.

In this paper, we report an application of the MPTM group as a promising 2'-hydroxyl protecting group to oligoribonucleotide synthesis.

RESULTS AND DISCUSSION

Treatment of 2'-O-(1,3-benzodithiol-2-yl)-3',5'-O-(1,1,3,3-tetraisopropyldisiloxan-1,3-diyl)uridine (**1**) with TBTH in the presence of AIBN in benzene followed by successive reaction with MeI gave a desired product (**2**) in 73% yield. Benzoylation of **2** followed by desilylation gave the compound (**3**) having the MPTM group at the 2'-oxygen. A monomer building block (**4**) of uridine was synthesized via a two-step reaction from **3**. A cytidine unit (**5**) was prepared from the same synthetic intermediate **2** by applying a procedure of uracile - cytosine conversion reported by Sung (3) and Reese.(4)

Compound **5** was treated by a phosphinate reagent in pyridine to give a phosphodiester component in quantitative yield. The selective removal of the dimethoxytrityl group from **4** was performed by 1% trifluoroacetic acid in dichloromethane to give the 5'-OH component without damage of other protecting groups. From these two component the protected dimer of CpUp was prepared. Similarly, a fully protected trimer block of CpUpG was synthesized by condensation between a OH component of guanosine and a 3'-terminal phosphodiester component of the prodected CpUp dimer. Deprotection of the fully protected trimer was performed in the usual way except for the final removal of the MPTM group which could be effectively eliminated by silver-ion catalyzed hydrolysis. The unprotected trimer was isolated and characterized by enzyme assay.

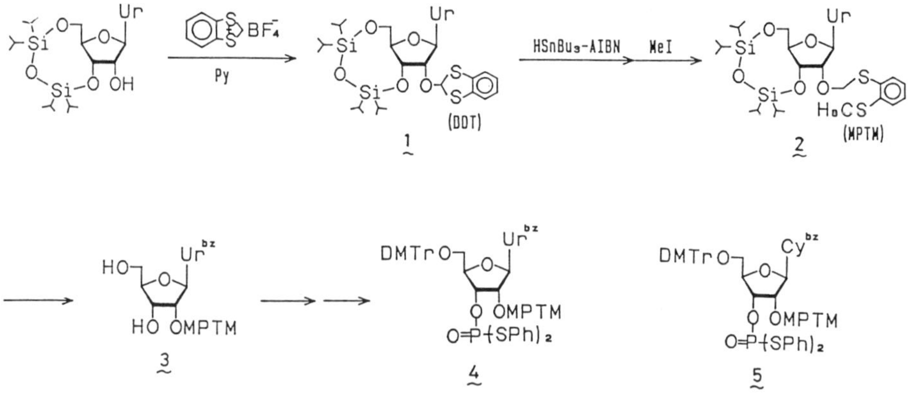

REFERENCES

1 Sekine, M. and Nakanishi, T. J. Org. Chem. in press.
2 Sekine, M. and Nakanishi, T. in preparation.
3 Sung, W. L. (1982) J. Org. Chem., 47, 3623-3628.
4 Divakar, K. J. and Reese, C. B. (1982) J. Chem. Soc., Perkin I, 1171-1176.

Syntheses and properties of oligodeoxyribonucleotide analogues having diastereochemically pure phosphoramidate linkages

Hiroaki Ozaki, Masanori Kitamura, Kazushige Yamana[1], and Takeo Shimidzu*

Division of Molecular Engineering, Graduate School of Engineering, Kyoto University, Sakyo-ku, Kyoto 606 and [1]Department of Applied Chemistry, Himeji Institute of Technology, 2167 Shosha, Himeji 671-22, Japan

ABSTRACT

The oligothymidilate analogues, having stereoregular and alternative phosphormorpholidate/phosphodiester backbone, were synthesized from the diastereochemically pure dimer blocks by phosphorbisamidite method. The phosphormorpholidate linkage of the oligothymidilate analogues were resistant to snake venom phosphodiesterase, and no cleavage of the phosphodiester linkage of the analogues were slowly cleaved by the nuclease was slow. The abilities of isomers to form the complexes with poly(dA) greatly depended on their structures.

INTRODUCTION

Oligonucleotide analogues containing modified backbone have been receiving an increasing attention because of their potential as gene regulatory substances[1]. These modifications result in the formation of diastereoisomers that differ in configuration about the phosphorus atom. Although several reports have shown that oligonucleotides possessing diastereochemically pure phosphorothioate or methylphosphonate linkages can be used for these studies[2,3], oligonucleotide having diastereochemically pure phosphoramidate linkage has not been used. We have already synthesized diastereochemically pure dinucleoside phosphormorpholidate derivatives[4] and characterized their derivatives using HPLC and NMR. In this paper, we describe the synthesis of oligothymidilate analogues having stereoregular alternating phosphormorpholidate/phosphodiester backbone and the preliminaly studies on the abilities to form the complex with poly(dA) and on the enzyme resistance.

RESULTS AND DISCUSSION

5'-O-Dimethoxytritylthymidylyl(3'-5')phosphormorpholyl 3'-O-benzoylthymidine was synthesized by the previously described procedure[4]. The isolation of chirally pure products was carried out by silica gel chromatography. After deprotection of 3'-O-benzoyl group, the

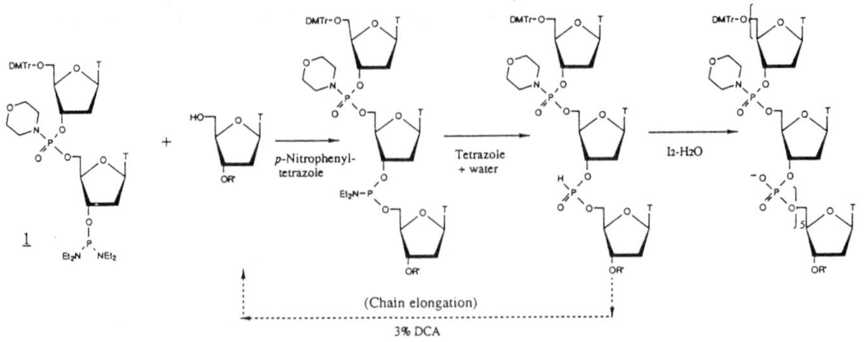

diastereochemically pure dithymidine phosphormorpholidate derivatives were phospholylated by bisdiethylaminophosphorchloridite. The obtained phosphorbisdiethylamidite derivatives **1** were used in solid-phase bisamidite method[5] to obtain the undecathymidilate analogues. The synthetic cycle contains a coupling reaction of 0.05M **1** and 0.1M p-nitrophenyltetrazole in dry CH_3CN, a hydrolysis reaction using 0.5M tetrazole in CH_3CN-H_2O(4:1,v/v), and a deprotection of dimetoxytrityl group using 3% dichloroacetic acid in CH_2Cl_2. After the end of cycle, the support-bound oligonucleotide analogue was oxidized with 0.1M I_2 in H_2O-lutidine-THF (1:2:2,v/v/v)(10 min). The products were removed from solid-suppot using conc. NH_4OH and purified by reversed-phase HPLC.

Treatment of each undecathymidilate analogue having diastereochemically pure phosphormorpholidate linkages with 10% isoamyl nitrite in H_2O-CH_3COOH-EtOH(1:4:5, v/v/v)(45°C, 45hr)[6] generated undecathymidine containing all phosphodiester linkages. The phosphormorpholidate of both undecathymidilate analogues were resistant to snake venom phosphodiesterase, and clevage of the phosphodiester linkage of the analogues by snake venom phosphodiesterase was slow. Furthermore, one of the two undecathymidilate analogues having diastereochemically pure phosphormorpholidate linkage did not form a duplex with poly(dA) under our experimental condition(150 mM sodium chloride/10 mM sodium phosphate buffer, pH 7.2, 5-80°C), while the other one formed a duplex with poly(dA) having almost similar Tm value as that of the native undecathymidine with poly(dA).

REFERENCES
1. Marcus-Sekura,C.J. (1988) Anal.Biochem., 172, 289.
2. Connolly,B.A., et al. (1984) Biochemistry, 23, 3443.
3. Miller, P.S., et al. (1980) J.Biol.Chem., 255, 9659.
4. Shimidzu,T., et al. (1988) Nucleic Acids Res., Symp.Series No.19, 1.
5. Yamana,K., et al. Tetrahedron, in press.
6. Froehler,B., et al. (1988) Nucleic Acids Res., 16, 4831.

Streptomyces ATP nucleotide 3′-pyrophosphokinase is also an ApppA nucleotide 2′, 3′-cyclic phosphokinase

Jun-Ichiro Mukai, Rokuya Ikuta, Toshiharu Shoji, and Akinori Hirashima[1]

Graduate School of Genetic Resources Technology and [1]Department of Agricultural Chemistry, Kyushu University, Fukuoka, Japan

Streptomyces nucleotide 3'-pyrophosphokinase(EC 2.7.6.4) transfers 5'-β,γ-pyrophosphoryl group of ATP to a variety of purine and pyrimidine nucleotides at their 3'-OH site. Among the interesting products of its action are ppGpp (a signal molecule of bacterial stringent control), ppApp (a bacterial sporulation factor), ppCpp stimulating mRNA-specific translation, NAD- and FAD-3'-pyrophosphates totally devoid of cofactor activity, CoA- and AcCoA-3'-pyrophosphates endowed with 2- and 6-fold enhanced cofactor and allosteric effector activity with phosphotransacetylase and phosphoenolpyruvate carboxylase respectively, etc(1). Little can be known, however, about cellular function(s) of the enzyme from above study. We now explored the enzyme action on several ATP-analogous diadenosine 5',5'-polyphosphates or known alarmones.

A mixture of 5 mM ApppA, 50 mM Tris-HCl, pH 9, 6 mM $CoCl_2$ and 90 U/ml enzyme was reacted at 37°C for 3 hr, and passed thru a borate gel column to be divided into two, namely break-thru and bound fractions. The former yielded isomeric AMP-2'- and 3'-phosphates on alkaline as well as acidic hydrolysis, and 2',3'-cyclic AMP on phosphomonoesterase digestion, but was resistant to 3'-specific nuclease P_1. Comparison with authentic sample, together with FAB mass spectral analysis giving an $(M + 1)^+$ ion peak at m/z 410, afforded its identification as AMP-2',3'-cyclic phosphate. The latter was identified as AMP. A similar kinase reaction with CpG added as a possible acceptor gave two fractions again on borate gel filtration. The break-thru contained AMP-2',3'-cyclic phosphate as before and another component. Its digestion with pancreas RNase gave Cp + G>p instead of G, thus identifying it as CpG>p (Fig. 1). We thus consider following reactions taking place, respectively.

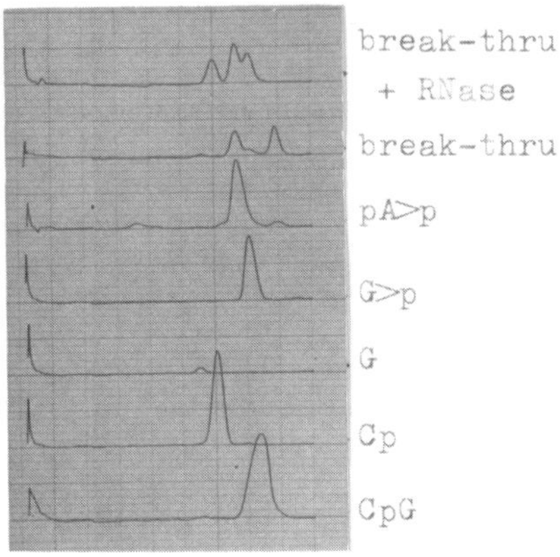

Fig. 1. Avicel thin layer chromatography was developed with isobutyric acid/H_2O/28 % ammonia (66/33/1) and scanning taken at 268/310 nm with a Shimadzu chromatoscanner CS 900.

ApppA -> pA + E-ppA -> pAppA -> pAp-E + pA -> pA>p + pA

ApppA + CpG -> pA + CpGppA -> 2 pA + CpGp-E -> 2 pA + CpG>p

Diadenosine tetra- and pentaphosphates were also reactive but diphosphate was not. These are under study.

ACKNOWLEDGEMENTS

We thank Sanraku-Ocean Co., Godo Shusei Co., and Meiji Seika Co. for bacterial cultures. We are also indebted to Mr. Isobe for mass spectral analysis.

REFERENCE

1. For review, see: Mukai, J.-I.(1988) In Kleinkauf, H., von Döhren, H. and Jaenicke, L. (ed.). The Roots of Modern Biochemistry. Walter de Gruyter & Co., N.Y. Berlin, pp 603-607.

Solid-phase synthesis of oligoribonucleotides

Yao-Zhong Xu, Christian Lehmann, George Slim, Chris Christodoulou, Zu-Kun Tan, and Michael J.Gait

Medical Research Council, Laboratory of Molecular Biology, Hills Road, Cambridge CB2 2QH, UK

ABSTRACT

An efficient method is described for solid-phase synthesis of oligoribonucleotides that involves use of the 9-fluorenylmethoxycarbonyl group (Fmoc) for 5'-protection, 4-methoxytetrahydropyran-4-yl (Mthp) for 2'-protection and a phosphoramidite coupling procedure.

INTRODUCTION

Synthetic oligoribonucleotides have important applications in the study of RNA structure, of RNA-protein interactions and of catalytic RNA. Such applications have prompted a search for a reliable solid-phase method for synthesis of oligoribonucleotides.

Of crucial importance is the choice of protecting groups for the 2'- and 5'-positions of nucleoside residues. 2'-Protecting groups must remain stable throughout oligonucleotide assembly and be removed at the end of synthesis by a procedure that does not cause isomerisation or cleavage of the phosphodiester linkage. 5'-Protecting groups must be removed at every cycle of oligonucleotide assembly without side reactions taking place.

One strategy is to use a conventional acid-labile protecting group, such as the dimethoxytrityl group, for the 5'-position in combination with either t-butyldimethylsilyl[1], o-nitrobenzyl[2], or the acetal CTMP[3] for the 2'-position. An alternative approach is to use a conventional acid-labile group, such as 4-methoxytetrahydropyran-4-yl[4], to protect the 2'-position and develop a new 5'-protecting group removable under mild non-acidic conditions. We favour this second approach since oligoribonucleotides can be stored indefinitely and safely in their 2'-protected form and, before use, can be deprotected under mild acidic conditions that do not give rise to isomerisation or chain cleavage.

For 5'-protection we have chosen the 9-fluorenylmethoxycarbonyl group (Fmoc)[5] and show that it is highly suitable for use in solid-phase assembly of oligoribonucleotides[6].

RESULTS AND DISCUSSION

Starting from 2'-O-methoxytetrahydropyranyl uridine, 4-N-benzoylcytidine, 6-N-benzoyladenosine or 2-N-isobutyrylguanosine, monomers suitable for solid-phase synthesis were prepared in two steps by 1) preparation of the corresponding 5'-O-Fmoc deravatives in 52-65% yields and 2) phosphitylation to give the corresponding 3'-O-(2-cyanoethyl-N,N-diisopropyl) phosphoramidites in 52-78% yield as recently described[6].

Nucleoside-functionalised supports were prepared by reaction of 5'-O-Fmoc-2'-O-methoxytetrahydropyranyl nucleoside 3'-O-succinates with sarcosine-functionalised CPG glass. The resultant linkage is stable to treatment with DBU but cleaved rapidly by concentrated ammonia at 50° [7].

Assembly of oligoribonucleotides is as shown (Figure). Repetitive yields, as judged by UV spectroscopic measurement at 305nm of cleaved Fmoc, are about 96%.

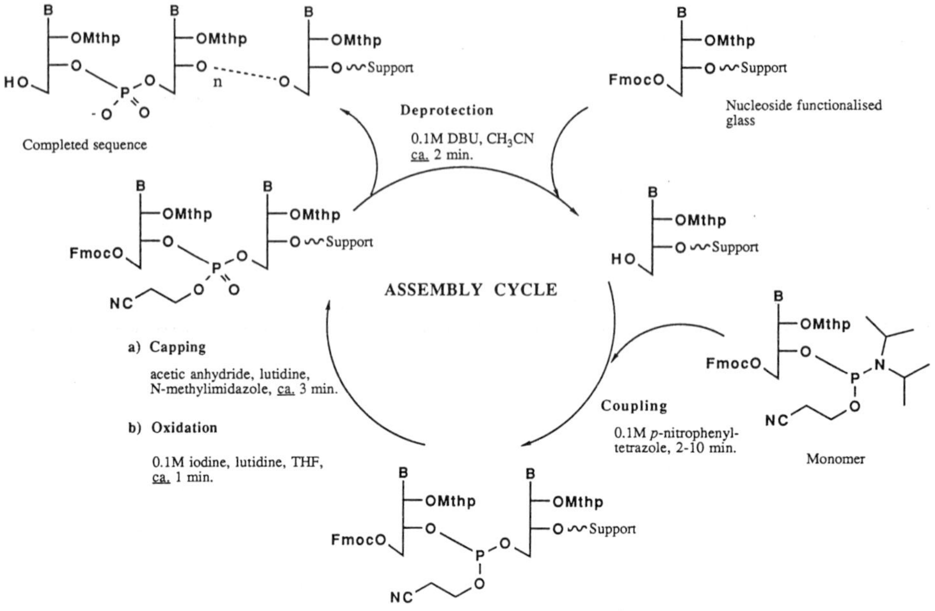

Completed sequences are released from the support and base-protecting groups are removed by treatment of the support with concentrated aqueous ammonia solution at 55° for 5 hours. Oligoribonucleotides, still containing 2'-protecting groups, are purified by reversed phase or ion-exchange h.p.l.c. and stored in 2'-protected form. Full deprotection is achieved by treatment with hydrochloric acid at pH 2 for 2-4 hours, neutralisation and oligoribonucleotides isolated in 80-90% yield by gel filtration.

Several oligoribonucleotides up to 20 residues in length have been prepared. Some of these are being used to study the structure and activity of artificial ribozymes.

REFERENCES

1) Usman, N., Ogilvie, K.K., Jiang, M.-Y. and Cedergren, R.J. (1987) J. Amer. Chem. Soc., **109**, 7845-7854.
2) Tanaka, T., Tamatsukuri, S. and Ikehara, M. (1986) Nucleic Acids Res., **14**, 6265-6275.
3) Rao, T.S., Reese, C.B., Serafinowska, H.T., Takaku, H. and Zappia, G. (1987) Tetrahedron Letters, **28**, 4897-4900.
4) Reese, C.B., Saffhill, R. and Sulston, J. (1967) J. Amer. Chem. Soc., **89**, 3366-3368.
5) Gioeli, C. and Chattopadhyaya, J.B. (1982) J. Chem. Soc. Chem. Commun. 672-674.
6) Lehmann, C., Xu, Y.-Z., Christodoulou, C., Tan, Z.-K. and Gait, M.J. (1989) Nucleic Acids Res., **17**, 2379-2390.
7) Brown, T., Pritchard, C. E., Turner, G. and Salisbury, S.A. (1989) J. Chem. Soc. Chem. Commun., in the press.

Synthesis of 8-diamino-substituted adenosine, inosine, and guanosine

J.A.Mauriņš, Dz.E.Zariņa, R.A.Paegle, and M.J.Līdaks

Institute of Organic Synthesis, Latvian SSR Academy of Sciences, Riga, USSR

ABSTRACT

Adenosine, inosine and guanosine derivatives were prepared, modified at the C-8 atom with the aid of diamines (1,3-diaminopropane, 1,4-diaminobutane and 1,5-diaminopentane).

DNA molecules containing a modified pyrimidine base-putrescinyl thymine have been isolated from bacteriophages [1]. It is believed that the presence of positively charged amino groups interacting with the phosphate groups is responsible for the high thermal stability of such DNA [2]. Recently, T.Ueda has reported the synthesis and properties of deoxyoligonucleotides containing putrescinyl thymine [3]. For further investigation of modified oligonucleotides, it would be advisable to synthesize molecules containing purine bases with a polyamine residue at C-8.

As the first step, we prepared 8-diamino-substituted nucleosides 2a-i by reacting 8-bromo derivatives of adenosine, inosine and guanosine with 1,3-diaminopropane, 1,4-diaminobutane (putrescine) and 1,5-diaminopentane (cadaverine).

8-(3-Aminopropylamino)adenosine 2a has been prepared by heating 1,3-diaminopropane and bromadenosine 1a in the presence of water [4,5]. We found that 8-diamino-substituted adenosines 2a-c could be obtained by boiling 8-bromoadenosine 1a with diamine in 96% ethanol. The highest yield of diaminonucleosides 2a-c and decreased formation of symmetrical nucleosides like compound 3, was attained in diamine excess (no less than 5 moles of amine per 1 mole of 8-bromadenosine).

More rigid conditions were required to prepare inosine and guanosine 8-diaminonucleosides 2d-i - the reaction was conducted in sealed ampoules at 120-130°C (inosine derivatives) and 130-140°C (guanosine derivatives).

Following purification by ion-exchange chromatography (Dowex H^+-form, gradient elution by 25% NH_4OH from 0% to 10%, v/v) and/or silica gel chromatography (eluent ethanol - 25% NH_4OH 4 : 1, v/v) the yield of nucleosides 2a-i amounted to 50-65%.

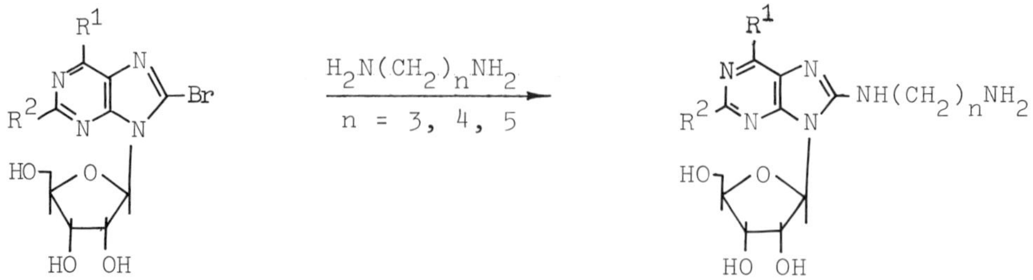

1 2

a, R^1 = NH_2; R^2 = H
b, R^1 = OH; R^2 = H
c, R^1 = OH; R^2 = NH_2

a, R^1 = NH_2; R^2 = H; n = 3
b, R^1 = NH_2; R^2 = H; n = 4
c, R^1 = NH_2; R^2 = H; n = 5
d, R^1 = OH; R^2 = H; n = 3
e, R^1 = OH; R^2 = H; n = 4
f, R^1 = OH; R^2 = H; n = 5
g, R^1 = OH; R^2 = NH_2; n = 3
h, R^1 = OH; R^2 = NH_2; n = 4
i, R^1 = OH; R^2 = NH_2; n = 5

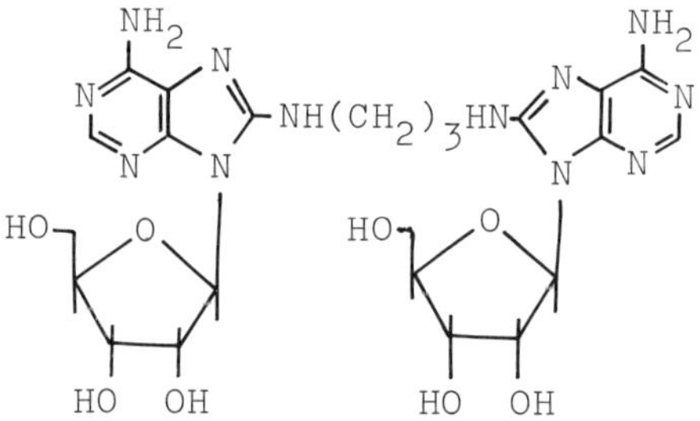

3

The compounds synthesized were inactive when assayed on melanoma B16 cells *in vitro*.

REFERENCES

1. Kropinski, A.M., Bose, R.J. and Warren, R.A.J. (1973) Biochemistry, 12, 151.
2. Gerhard, B. and Warren, R.A.J. (1982) Biochemistry, 21, 5458.
3. Takeda, T., Ikeda, K., Mizuno, Y. and Ueda, T. (1987) Chem. Pharm. Bull. 35, 3558.
4. Kajander, E.O. and Raina, A.M. (1981) Biochem. J., 193, 503.
5. Holy, A. (1983) Coll. Czech. Chem. Commun., 48, 1910.

The introduction of reporter groups at multiple and/or specific sites in DNA containing phosphorothioate diesters

Nancy E.Conway, Jacqueline Fidanza, and Larry W.McLaughlin*

Department of Chemistry, Boston College, Chestnut Hill, MA 02167, USA

*Author to whom correspondence should be directed

ABSTRACT

DNA fragments containing phosphorothioate diesters provide nucleophilic sites which are amenable to labeling by spin labels or fluorophores. Selecting the position for an individual phosphorothioate diester allows highly specific placement of the reporter group. The substitution of a phosphorothioate diester for each and every internucleotidic phosphodiester allows the incorporation of multiple reporter groups; ideally one for each nucleoside residue. With the use of multiple fluorophores a post-assay fluorescent labeling technique has been developed which allows the detection of DNA fragments with the "naked-eye" in the low femtomolar (10^{-15} moles) range.

INTRODUCTION

Phosphorothioate diesters can be easily incorporated into DNA fragments by chemical (1,2) and enzymatic (3) methods. The resulting phosphorothioate containing nucleic acids have been used to study the stereochemical course of restriction endonucleases (2), to protect against nuclease digestion (4,5), in a program for oligonucleotide-directed mutagenesis (6) and in degradation reactions which can be exploited for DNA sequencing (7,8).

We have observed that the phosphorothioate diester can also function as a site for modification by reporter groups such as the fluorophore monobromobimane (9). The labeled phosphorothioate triesters are stable at neutral and acidic pH values but hydrolyze (with P-S bond cleavage) under alkaline conditions. In the present work we have used either a spin label or fluorophores to modify phosphorothioate-containing DNA at single or multiple positions.

RESULTS AND DISCUSSION

The dodecamer, d(CGCA(s)AAAAAGCG), could be reacted with the fluorophore monobromobimane or the spin label 3-(2-iodoacetamido)-PROXYL and either product isolated by HPLC. Using the complementary strand of DNA, d(CGCTTTTTTGCG), the stability of the double stranded dodecamer containing a single reporter group bound to an internucleotidic phosphorothioate diester was examined. The Tm value obtained (55 °C) for the labeled double stranded dodecamer did not differ significantly from the unlabeled duplex. This suggests that the placement of a reporter group at the internucleotidic phosphorus does not disturb the native DNA helical structure and can be used effectively to monitor nucleic acid dynamics and structural modulation.

Multiple labeling experiments have been directed toward the detection of nucleic acid fragments via emitted fluorescence. In this case, DNA fragments have been generated with multiple phosphorothioate diesters. After resolution by polyacrylamide gel electrophoresis, monobromobimane can be introduced to the gel to detect the DNA fragments by "post-assay" labeling. After diffusion through the gel, multiple bimane fluorophores become covalently attached to the DNA at the phosphorothioate diesters (Scheme 1), while the fragments are still embedded within the gel matrix. The success of this approach relies upon the observation that the bimane labeled phosphorothioates are highly fluorescent while monobromobimane is itself largely non-fluorescent (10).

Scheme 1 "Post-Assay" Labeling of Phosphorothioate-Containing DNA Fragments

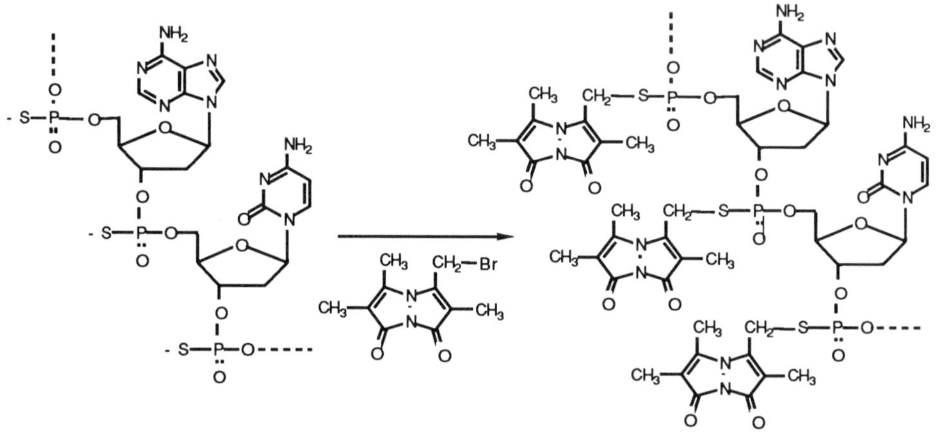

A 444-mer DNA fragment was generated by Ava I digestion of the RF form of M13 DNA prepared by using a mixture of 2-deoxynucleotide-α-thiotriphosphates and native triphosphates. The 444-mer fragment could be prepared with 104 (dATPαS), 223 (dATPαS, dTTPαS) or 328 (dATPαS, dTTPαS, dGTPαS) phosphorothioate diesters. After post-assay labeling, the limit of detection to the "naked-eye" was 41, 27 and 14 fmol (10^{-15} moles) respectively.

ACKNOWLEDGEMENT This work was supported by an NIH grant (GM37065) to L.W.M.

REFERENCES

1. Connolly, B.A., Potter, B.V.L., Eckstein, F., Pingoud, A. and Grotjahn, L. (1984) *Biochemistry* **23**, 3443-3453.
2. Stec, W. J., Zon, G., Egan, W. and Stec, B. (1984) *J.Amer.Chem.Soc.* **106**, 6077-6079.
3. Potter, B. V. L. and Eckstein, F. (1984) *J.Biol.Chem.* **259**, 14243-14248.
4. Putney, S., Benkovic, S. and Schimmel, P. (1981) *Proc.Natl.Acad.Sci.USA* **78**, 7350-7354.
5 Jasin, M., Regan, L., and Schimmel, P. (1983) *Nature* **306**, 441-447.
6. Taylor, J.W., Schmidt, W., Cosstick, R., Okruszek, A. and Eckstein, F. (1985) *Nucl. Acids Res.* **13**, 8749-8764.
7. Gish, G. and Eckstein, F. (1988) *Science* **240**, 1520-1522.
8. Nakamaye, K. L., Gish, G., and Eckstein, F. (1988) *Nucl. Acids Res.* **16**, 9947-9959.
9. Hodges, R. , Conway, N.E. and McLaughlin, L.W. (1989) *Biochemistry* **28**, 261-267.
10. Kosower, E.M. and Pazhenchevsky, B. (1980) *J.Amer.Chem.Soc.* **102**, 4983-4993.

Labile protecting groups in tRNA synthesis

C.Chaix, A.M.Duplaa, D.Gasparutto, D.Molko*, and R.Téoule

Département de Recherche Fondamentale, Laboratoires de Chimie, Centre d'Etudes Nucléaires de Grenoble, 85X 38041 Grenoble Cédex, France

ABSTRACT The use of labile protecting groups for the protection of the exocyclic amino function of adenine, guanine and cytosine has two main advantages in RNA synthesis. Final deprotection in concentrated aqueous ammonia takes place in milder conditions which are more compatible with the sensitivity of oligoribonucleotides towards alkali-conditions. The introduction of fragile bases such as certain modified bases encountered in the primary structure of t-RNA is feasible. The chemical synthesis of RNA fragments constituting the primary structure of *B.subtilis* f-methionine t-RNA is described.

The choice of appropriate protecting groups for the exocyclic amino-functions of nucleic bases is a key step in oligoribonucleotide synthesis. RNA fragments are very sensitive to alkaline hydrolysis and the modified nucleosides constituting a part of the primary structure of transfert ribonucleic acids are frequently alkali-sensitive. The classical protecting groups introduced by Khorana (1) necessitate a drastic treatment in concentrated aqueous ammonia for 17 hours at 60° C for their complete removal. These conditions cancel any hope of synthesizing oligoribonucleotides incorporating 5,6-dihydrouridine or pseudo-uridine (DHU) which are modified nucleosides encountered in most transfert-RNAs. Phenoxyacetyl group (for the protection of guanine and adenine) and isobutyryl or acetyl group (for cytosine) has been sucessfully employed both in DNA (2) and RNA (3,4) synthesis. We present here their use for the synthesis of RNA fragments bearing fragile nucleosides such as 5,6-dihydrouridine and pseudo-uridine.

RESULTS AND DISCUSSION

Preliminary studies were necessary to test the stability of DHU and pseudo-uridine in the oligonucleotide synthesis conditions. These investigation were made by reversed phase HPLC analysis of crude reaction mixtures of the modified nucleosides, treated in concentrated aqueous ammonia or detritylation conditions respectively.

DHU showed less than 5 % degradation in 4 hours at room temperature in ammonia and was stable in acidic medium, indicating that it is suitable for incorporation in oligonucleotides, using the labile base protecting groups.

Pseudo-uridine was more stable in ammonia. It was isomerized in the range of 7 % in a mixture of α and β furanose and pyranose forms after 20 hours in 80 % acetic acid at 65° C. This low level of degradation should not occur during the assembling step of the RNA fragments.

The protection of DHU, pseudo-uridine and 5-methyl uridine (ribosyl thymidine) was made by a 4,4'-dimethoxytrityl group on the 5' position and a tertiobutyldimethylsilyl group on the 3' position. The corresponding phosphoroamidite monomers were prepared according to reported procedure (4).

These monomers were used in cyanoethylphosphoroamidite technology (5) to synthesize various oligonucleotides up to 42-mer constituting the sequence of *B. subtilis* f-methionine t-RNA. These RNA fragments exhibited a correct migration shift upon PAGE. Their complete characterization is under way.

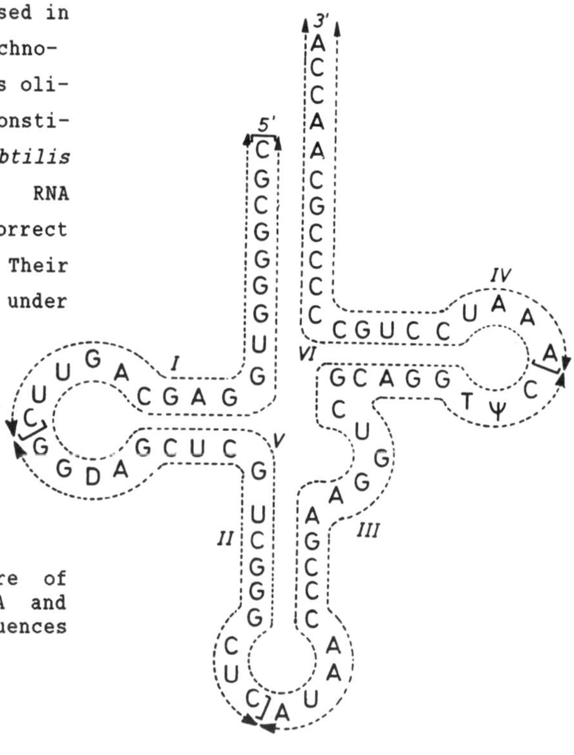

Figure 1: Primary structure of *B.subtilis* f-methionine t-RNA and numbering of the sequences mentionned in this paper.

REFERENCES

1 Schaller H., Weimann G., Lerch B. & Khorana H.G. (1963) J.Amer.Chem.Soc. 85,3821.
2 Schulhof J.C., Molko D. & Téoule R. (1987) Nucleic Acids Res. 15,397.
3 Wu T., Ogilvie K.K. & Pon R.T. (1988) Tet.Lett. 29,4249.
4 Chaix C., Molko D. & Téoule R. (1989) Tet.Lett. 30,71.
5 Sinha N.D., Biernat J., Mc Namus J. & Köster H. (1984) Nucleic Acids Res. 12,4539.

Recent studies on nucleoside H-phosphonothioate and nucleoside methylphosphonothioate synthesis

Jacek Stawiński*, Tomas Szabó, Mats Thelin, Erik Westman, and Rula Zain

Department of Organic Chemistry, Arrhenius Laboratory, University of Stockholm, 106 91 Stockholm, Sweden

Abstract
The reaction of suitably protected nucleosides with phosphinic acid in the presence of condensing agents has been investigated and applied for the preparation of nucleoside 3′-H-phosphonothioates. Studies on synthesis and separation of diastereoisomers of nucleoside 3′-methylphosphonothioates are also discussed.

As a continuation of our studies on the H-phosphonate chemistry, we have become interested in investigation of chemical properties and possible synthetic applications of nucleoside H-phosphonothioate esters.

At the early stage of this investigation the main obstacle was lack of a suitable method for the preparation of nucleoside H-phosphonothioates. In 1965, Michalski *et al.*[1] described a method for the preparation of dialkyl H-phosphonothioates, which recently has been adapted for nucleoside H-phosphonothioate diesters by Caruthers *et al.*[2] Even if ethyl H-phosphonothioate can be prepared *via* hydrolysis of the appropriate diester[1], such an approach seemed not to be suitable for the synthesis of nucleoside H-phosphonothioate monoesters. Thus, we decided to explore a possibility of synthesizing the compounds in question *via* phosphinate intermediates as shown below:

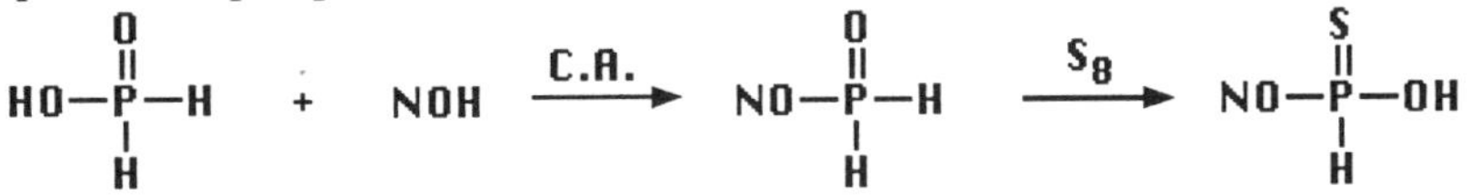

The reaction proved to be more complicated than the scheme might suggest. When 3 equiv. of phosphinic acid and 3 equiv. of pivaloyl chloride in pyridine were used, TLC analysis shoved rather fast disappearance of the starting nucleoside and formation of a new, apparently neutral compound. This, however, was not the desired nucleoside phosphinate but a symmetrical dinucleoside H-phosphonate diester. More detail studies revealed that this probably was a consequence of disproportionization of phosphinic acid under the reaction conditions which resulted in oxidation of nucleoside phosphinate and its subsequent coupling with a nucleoside to form the H-phosphonate diester. The ratio of nucleoside to phosphinic acid was found to be a critical factor in that reaction. When this was 1.5-2, the nucleoside phosphinate was formed as a single nucleotidic product and it was converted into the nucleoside H-phosphonothioate by *in situ* oxidation with sulfur.

Several deoxyribo- and ribonucleoside H-phosphonothioates have been synthesized in this way in overall yield of ca 90%. Preliminary studies concerning the reactivity of nucleoside H-phosphonothioates during formation of H-phosphonothioate diesters have been recently reported[3].

Another part of our investigations dealt with nucleoside methylphosphonothioates synthesis. Oligonucleotides analogues containing methylphosphonothioate residues are gaining interest as a potential antiviral drugs. An interesting feature of methylphosphonothioates is chirality of a phosphorus centre at the mononucleotide level, which make it possible to investigate stereospecific synthesis of methylphosphonothioate diesters.

To this end, separate diastereoisomers of 5´-O-dimethoxytritylthymidine 3´-methylphosphonothioate have been obtained using procedure as on the scheme below:

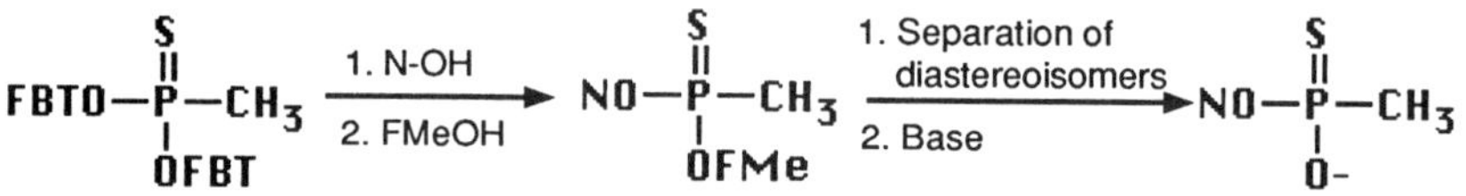

FBT = 6-trifluoromethybenzotriazole; FMeOH = 9´-fluorenylmethanol

9-Fluorenemethyl group was chosen for transient protection since it gave best separation of the diastereoisomers and it was easily removed using triethylamine or DBU in pyridine.

To succeed in the stereospecific synthesis of methylphosphonothioate diesters, the two reaction steps involved in an internucleotidic linkage formation, namely activation of a nucleoside methylphosphonothioate with a condensing agent and the reaction of the activated species with a nucleoside, have to be stereospecific.

Preliminary experiments concerning activation of methylphosphonothioate monoesters have shown that it is possible to do activation in a stereospecific manner, at least with some condensing reagents, e.g. chlorophosphates. Other reagents, e.g. arene sulfonic acid derivatives, which have a tendency to form pyromethylphosphonothioates from nucleoside methylphosphonothioates, seem to be unsuitable as activators in an attempted stereospecific synthesis of methylphosphonothioate diesters.

Acknowledgements

We are indebted to Prof. Per J. Garegg for his interest, to the National Swedish Board for Technical Development and the Swedish Natural Science Research Council for financial support.

REFERENCES

1. J. Michalski, Z. Tulimowski, R. Wolf, Ber., **98**, 3006-3009 (1965).
2. J. Nielson, W.K.-D. Brill, M. Caruthers, Tetrahedron Lett., **29**, 2911-2914 (1988).
3. J. Stawinski, M. Thelin, R. Zain, Tetrahedron Lett., (1989), in press.

Antiviral activities and the mechanism of 9-β-D-ribofuranosyl-6-alkylthiopurines on several RNA viruses from animals

Mineo Saneyoshi, Masahide Hasobe, and Kiichi Yamamoto*

Faculty of Pharmaceutical Sciences, Hokkaido University, Sapporo 060 and Department of Virology and Rickettsiology, National Institute of Health, Tokyo 141, Japan

ABSTRACT

A series of 9-β-D-ribofuranosyl-6-alkylthiopurines (6-alkyl TI) were found to inhibit *in vitro* replication of infectious hematopoietic necrosis virus (IHNV), human influenza virus (IFV) and respiratory syncitial virus (RSV) with IC_{50} values of about 0.06 μg/ml, 0.7-1.5 μg/ml and 1-3 μg/ml, respectively. Viral RNA synthesis in infected cells in the presence of actinomycin D was inhibited by treatment with the compounds dose-dependently. It was also found that the decrease of rNTP pool size in infected cells was remarkably dose-dependent. From these findings, the mode of antiviral action of these compounds may be explained by rNTP inbalance in the treated group.

INTRODUCTION

Antiviral agents against herpes simplex type 1 (HSV-1) and valicella-zoster virus (VZV) have been successfully developed in recent year. Among them, Acyclovir (acycloguanosine) has been developed in clinical use in various countries including Japan. However, antiviral chemotherapy against RNA virus infections is still dificult in the present stage. We have previously developed a convenient system for evaluation of potential anti-RNA viral agents on microtiter plate with 96 wells (1-2). Infectious hematopoietic necrosis virus (IHNV)-CHSE 214 cells combination is used as a model animal virus probe for the evaluation of compounds. This method was successfully applied to antiviral evaluations of unique antibiotic guanine 7-oxide in this laboratory(3-4). Using this newly developed method, several-sulfur containing purine nucleosides were found to have antiviral activity. We therefore, examined the structure-antiviral activity relations and their mechanism oof antiviral action; the results are described in this paper.

MATERIALS AND METHODS

Cells and Viruses : Host cells and RNA viruses used in this study were previously reported; IHNV and CHSE-214 cells (1-4), several strains of IFV with corresponding host cells, RSV and host cells are used as our previous paper (5).

Nucleoside analogues

The compounds used here were chemically synthesized by using the known method.

Assay of Antiviral Activity

CPE-spot reduction assay was employed in IHNV series (2-3). Virus plaque reduction method was used in the case of IFV and RSV (5).

RESULTS AND DISCUSSION

Antiviral Activities

Antiviral activity of 9-β-D-ribofuranosyl-6-alkylthiopurines against IHNV was examined first. According to the result, in vitro anti-IHNV activity of 10 kinds of nucleoside analogues are all active against this system. Among them, 6-methylthio- and 6-ethylthio- derivatives, were showed almost same anti-IHNV ability (IC_{50} = 0.06 μg - 0.07 μg/ml) and the IC_{50}value decreased as the carbon-chain length increased at 6-position of the analogues. However, for the ratio of antiviral effect in vitro versus cytotoxycity against uninfected host cells, 6-ethylthio derivatives was shown to have best score when compared with that 6-methylthio analogues. The result was confirmed in both series of IFV and RSV.

Mechanism of Antiviral Action

The viral RNA synthesis in virus-infected cells was dose-dependently inhibited with low concentrations of the analogues, which accompanied comparable decrease of viral CPE spot. On the other hand, HPLC analyses of cellular rNTP pool in the infected cells revealed a remarkable increase by 200-400% upon IHNV infection. The results clearly demonstrates that the high level of rNTP pool size are required for the viral replication in host cells. When infected cells treated with the analogues; CPE-spots caused by viral infection decreases dose-dependently and the rNTP pool also decreases to the non-infected cell level. Thus, the antiviral mechanism of the compound may be explained by rNTP inbalance in the treated group.

REFERENCES

1. Hasobe, M. and Saneyoshi, M. (1985) Bull. Jpn. Soc. Sci. Fish., 51, 1079-1084.
2. Hasobe, M. and Saneyoshi, M. (1985) Fish Pathol., 20, 343-351.
3. Hasobe, M., Saneyoshi, M., and Isono, K. (1985) J. Antibiotics, 38, 1581-1587.
4. Hasobe, M., Saneyoshi, M., and Isono, K. (1986) J. Antibiotics, 39, 1291-1297.
5. Yamamoto, K., Hasobe, M., and Saneyoshi, M. (1988) Acta Virol., 32. 386-392.

DNA strand scission by antitumor antibiotics. Detection of C-4′-hydroxylated abasic sites in DNA

Hiroshi Sugiyama, Hiroshi Kawabata, Tsuyoshi Fujiwara, Yukihiro Dannoue, Takahisa Tashiro, Teruo Matsuura, and Isao Saito

Department of Synthetic Chemistry, Faculty of Engineering, Kyoto University, Kyoto 606, Japan

ABSTRACT

Detection of C-4'-hydroxylated abasic site in calf thymus DNA was investigated. Upon heating with neutral hydrazine (90°C, 5 min) C-4'-hydroxylated abasic site formed by photo-excited Co(III)bleomycin was converted to the fragments having 3'-phosphoro-3"-pyridazinylmethylate as illustrated on Scheme 1. Subsequent enzymatic digestion of the reaction mixture provided four kinds of pyridazine mononucleotides (**1**, **2**, **3**, and **4**). The fact that the amount of free bases released from calf thymus DNA corresponded well to the amount of pyridazines indicates this reaction can be used for detection of C-4'-hydroxylated abasic site in DNA. By this assay, we first identified the formation of C-4'-hydroxy abasic sites in calf thymus DNA by neocarzinostatin.

INTRODUCTION

The chemistry of DNA backbone oxidation mediated by antitumor antibiotics is a topic of intense current interest. Bleomycins (BLM) and neocarzinostatin (NCS) are antitumor antibiotics which are believe to act tumor cell by breaking DNA. Fe(II)BLM + O_2 has been shown to cleave DNA into fragments which has a glicolic acid at 3' phostphate termini. Another lesion which is formed by Fe(II)BLM + O_2 is a aklali labile site, and its detail structure was identified as C-4'-hydroxylated abasic site by the use of d(CGCTTTAAAGCG).[1,2] This method was successfully applied to investigate the chemistry of strand scission induced by photo-excited Co(III)BLM. Here we report the identification of C-4'-hydroxylated abasic site formed in longer substrate, such as calf thymus DNA.

RESULTS AND DISCUSSION

It has been found that Fe(II)BLM + O_2 induces C-4'-hydroxylation of deoxyribose in deoxydodecanucleotide, d(CGCTTTAAAGCG) and upon heating with neutral hydrazine (90°C, 5 min), modified oligonucleotide was cleaved to generate 3'-phosphoro-3"-pyridazinylmethylate termini at 3'-end.[2] Since this transformation to pyridazine is specific and efficient for C-4'-hydroxylated abasic site in deoxyoligonucleotides, we decided to utilize this reaction to detect these sites in DNA. The strategy used to detect C-4'-hydroxyabasic site in DNA is outlined in Scheme 1. We first examined a detection of C-4'-hydroxylated abasic site in calf thymus DNA. The reaction mixture contained 1 mM (base concentration) calf thymus DNA and 300 μM Co(III)BLM was irradiated (366 nm, 1 h) and then treated with hydrazine (0.1 M NH2NH2-HCl, pH 7, 90°C, 5 min). After enzymatic digestion of the reaction mixture, formation of four nucleotide pyridazines (**1**, **2**, **3**, and **4**) together with unreacted nucleosides was clearly observed on reverse phase HPLC. The fact that amount of free bases released from calf thymus DNA corresponded well to the amount of pyridazines indicates this reaction can be used for detection of C-4'-hydroxylated abasic site in DNA.

Next investigated was the formation of pyridazine nucleotides from NCS treated DNA, since we have already found that d(CGTACG) was attacked by NCS at thymine3 to produce C-4'-hydroxy abasic site. When a calf thymus DNA was treated with NCS in the presence of 4-hydroxythiophenol, free bases were released. Subsequent hydrazine treatment of the reaction mixture and following enzymatic digestion, four pyridazine nucelotides (**1**, **2**, **3**, and **4**) were formed. These results clearly demonstrate that NCS oxidizes C-4' carbon to produce C-4' hydroxylated abasic site in DNA. This specific detection of 4'-hydroxy abasic site in DNA is very efficient and convenient, it may be quite effective to detect these sites in chemically altered DNA.

Scheme 1

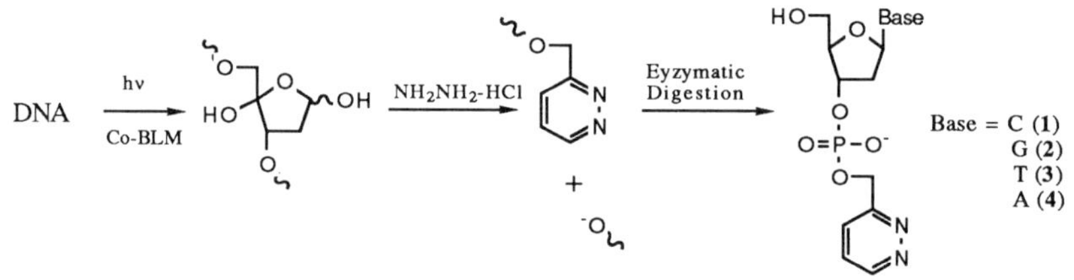

REFERENCES

(1) Sugiyama, H., Xu, C., Murugesan, N., Hecht, S. M. J. Am. Chem. Soc. 1985, 107, 4104

(2) Sugiyama, H., Xu, C., Murugesan, N.,Hecht, S. M. van der Marel, G. A. van Boom, J. H., Biochemistry 1988, 27, 58

DNA binding and cleavage by enediyne antitumor antibiotics, esperamicin and dynemicin

Takashi Shiraki and Yukio Sugiura

Institute for Chemical Research, Kyoto University, Uji, Kyoto 611, Japan

ABSTRACT Esperamicin A_1 and dynemicin represent a new class of DNA binding/damaging agents characterized by unique ten-membered ene-diyne core. The former antitumor antibiotic shows high DNA cleavage activity in the presence of dithiothreitol, whereas the latter in the presence of NADPH. Esperamicin A_1 favorably breaks T and C bases at oligopyrimidine regions such as 5'-CTC, 5'-TTC, and 5'-TTT sequences. Dynemicin attacks the bases at 3'-side of purine residues such as 5'-AG, 5'-AT, and 5'-GC sequences. These drugs also present good circumstantial evidence for minor groove interaction of B-DNA.

Esperamicin A_1 and dynemicin which were isolated from fermentations of *Actinomadura verrucosospora* or *Micromonospora chersina* are potent antitumor antibiotics. These compounds are representative of a new class of DNA binding and damaging agents. As well as calicheamicin γ_1, esperamicin A_1 and dynemicin contain several unusual structural features. In particular, the antibiotics are characterized by unique ten-membered ene-diyne core that produces a 1,4-dehydrobenzene-diradical by aromatization of the diyne-ene moiety. It is believed that this diradical species is responsible for the DNA damaging properties and cytotoxicity of these antitumor antibiotics.

The DNA cleavage activity of esperamicin A_1 remarkably increases in the presence of thiol compounds. The most preferential cutting sites of esperamicin are at deoxythymidylic acid residues, and the present frequency($T \geqslant C \gg A > G$) of bases attacked is clearly different from that($T > A \gg C > G$) of neocarzinostatin which is also one of the enediyne antibiotics. Esperamicin A_1 preferentially attacks T and C bases at oligopyrimidine regions such as 5'-CTC, 5'-TTC, and 5'-TTT. The 5'-<u>T</u>G and 5'-<u>C</u>G sequences are also preferred sites in the esperamicin-mediated DNA degradation.

Dynemicin showed high DNA cleavage activity in the presence of NADPH. In the dynemicin-mediated DNA strand scission, the considerably preferred cutting sites were the bases at 3'-side of purine residues such as 5'-A$\underline{G}$, 5'-A$\underline{T}$, and 5'-G$\underline{C}$ sequences.

The pre-treatment of aflatoxin B_1 to DNA gave no remarkable alterations on DNA cutting mode by dynemicin(Fig. 1). On the other hand, DNA cleavage patterns of esperamicin A_1 and dynemicin were significantly affected by the pre-treatment of DNA with netropsin and distamycin A, indicating that interaction of these antibiotics occurs through the minor groove of B-DNA.

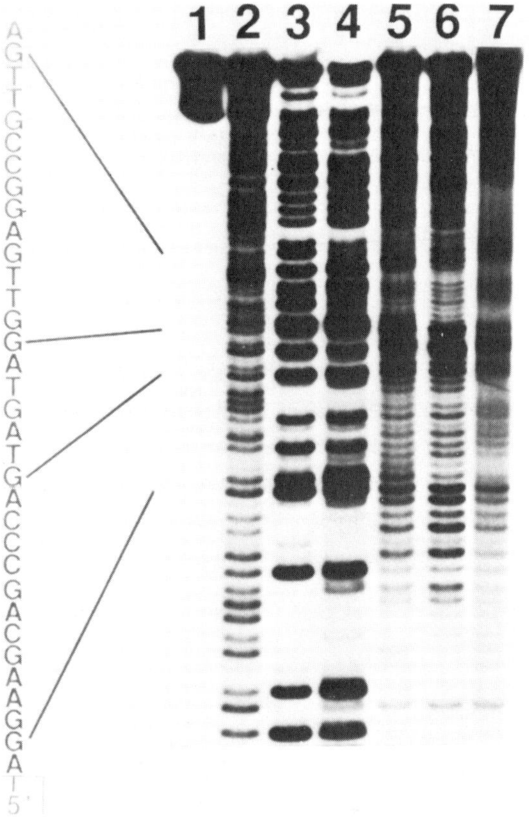

Fig. 1 DNA cutting modes by dynemicin in the pre-treatment of aflatoxin B_1(lane 5) or cis-dichlorodiammine Pt(II)(lane 7). Lanes 4 and 6 present the product of piperidine-treatment of DNA incubated with aflatoxin B_1 and the dynemicin-mediated cleavage for intact DNA. Lanes 1, 2, and 3 show DNA only, C+T, and G, respectively.

Synthesis of dinucleoside phosphorodithioates from nucleoside phosphonodithioates

Geoffrey M.Porritt and Colin B.Reese

Department of Chemistry, King's College London, Strand, London WC2R 2LS, UK

ABSTRACT

The conversion, in good yield, of 5'-O-(9-phenylxanthen-9-yl) thymidine (1) into the triethylammonium salt of its 3'-phosphonodithioate (2a) is described. The latter compound (2a) is converted into a dinucleoside phosphonothioate (4a), and thence into a corresponding dinucleoside phosphorodithioate (6a) and phosphorothioate (6;X=O).

INTRODUCTION

Unlike phosphorothioate analogues[1] of oligonucleotides, the corresponding phosphorodithioate analogues do not contain chiral internucleotide linkages and are resistant to phosphodiesterase-promoted digestion. For these reasons an interest has developed recently [2,3,4] in the preparation and properties of the latter analogues. The synthetic methodology used in the work published so far has been based mainly on the phosphoramidite approach[5] to oligonucleotide synthesis. We have developed a very convenient approach to the synthesis of phosphorodithioate analogues which involves the use of protected nucleoside 3'-phosphonodithioates [e.g. (2a)] as starting materials.

RESULTS

The synthetic procedures which were followed are indicated in the Scheme. The triethylammonium salt of 5'-O-(9-phenylxanthen-9-yl)thymidine 3'-phosphonodithioate (2a) [δ_P ($CDCl_3$) 85.0 (J_{PH} 546.3 Hz)] was isolated as a colourless solid in 75% yield; (4a) [δ_P ($CDCl_3$) 70.8 (J_{PH} 669.8 Hz), 72.2 (J_{PH} 664.0 Hz)] was obtained in 60% yield; (5a) [δ_P 113.2] was isolated as a

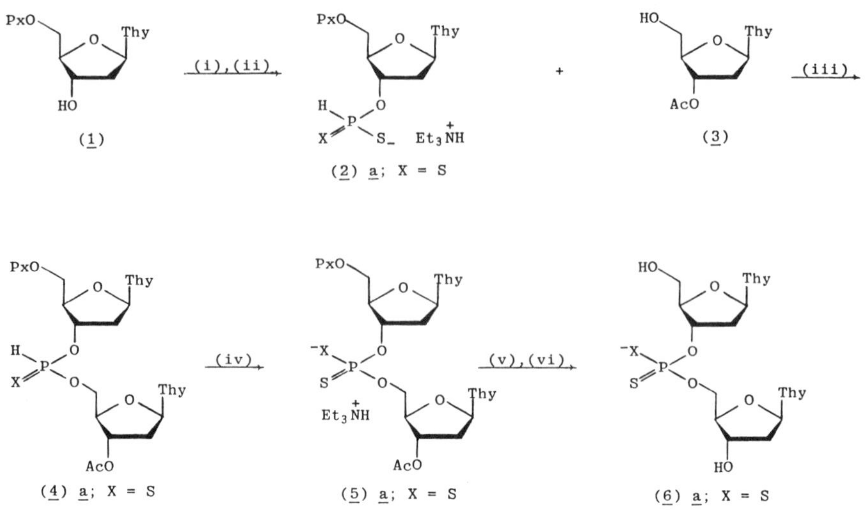

Reagents: (i) reagent derived from PCl_3 [4 mol. equiv. with respect to (1)], 1,2,4-triazole [12 mol. equiv.], Et_3N [13 mol. equiv.], tetrahydrofuran, N_2, -35°C, 15 min; (ii)(a) H_2S, Et_3N, tetrahydrofuran, RT, 30 min, (b) N_2, RT, 1 hr; (iii) Me_3CCOCl, C_5H_5N-MeCN (5:1 v/v), RT, 20 min; (iv) S_8, pyridine, RT, 15 min; (v) aqueous NH_3 (*d* 0.88), RT, 2.5 hr; (vi) 0.01M hydrochloric acid (pH 2), RT, 1 hr.

colourless solid in 72% yield, based on (3); (6a) [$\delta_P(D_2O)$ 114.1] was isolated as an HPLC-homogeneous compound. When (4a) was treated with 1.3 mol.equiv. of iodine in tetrahydrofuran - pyridine - water - 1-methyl-imidazole (40:5:4:1, v/v) and the protecting groups removed from the terminal hydroxy functions, a mixture of thymidylyl-(3'→5')-thymidine (ca. 4%) and the corresponding phosphorothioate [(6;X=O); ca. 96%] was obtained.

ACKNOWLEDGEMENT

We thank the Imperial Cancer Research Fund for financial support.

REFERENCES

1. Eckstein, F. (1985) Ann.Rev.Biochem. 54, 367-402.
2. Nielsen, J., Brill, W.K.-D., and Caruthers, M.H. (1988) Tetrahedron Lett. 29, 2911-2914; Brill, W.K.-D., Nielsen, J., and Caruthers, M.H. (1988) Ibid. 29, 5317-5320; Grandas, A., Marshall, W.S., Nielsen, J., and Caruthers, M.H. (1989) Ibid. 30, 543-546; Brill, W.K.-D., Tang, J.-Y., Ma, Y.-X., and Caruthers, M.H. (1989) J.Am.Chem.Soc. 111, 2321-2322.
3. Farschtschi, N. and Gorenstein, D.G. (1988) Tetrahedron Lett. 29, 6843-6846.
4. Stawinski, J., Thelin, M., and Zain, R. (1989) Tetrahedron Lett. 30, 2157-2160.
5. Beaucage, S.L. and Caruthers, M.H. (1981) Tetrahedron Lett. 22, 1859-1862.

Binding modes of inhibitors to ribonuclease T_1 as elucidated by the analysis of two-dimensional NMR

Ichio Shimada and Fuyuhiko Inagaki

Department of Molecular Physiology, The Tokyo Metropolitan Institute of Medical Science, 18−22 Honkomagome 3-chome, Bunkyo-ku, Tokyo 113, Japan

Abstract

Aromatic proton and high field shifted methyl proton resonances of RNase T_1 complexed with Guo, 2'GMP, 3'GMP or 5'GMP were assigned to specific amino acid residues by 2D-NMR spectra in comparison with the crystal structure of RNase T_1-2'GMP complex. The spatial proximities of amino acid residues as elucidated by NOESY spectra were found to be quite similar among free RNase T_1 and the inhibitor complexes, showing that large conformational changes did not occur upon complex formation. However, small but appreciable confomational changes were induced which were reflected by the systematic chemical shift changes of some amino acid residues in the active site. Furthermore, we confirmed that RNase T_1 contains two specific binding sites, one for the guanine base and the other for the phosphate moiety. The inhibitors are forced to adapt their conformations to fit the guanine base and the phosphate moiety to each binding site on the enzyme. This is consistent with our previous studies that 2'GMP and 3'GMP take *syn* form as a bound conformation, while 5'GMP takes *anti* conformation around glycosidic bonds .

Results

RNase T_1 contains three histidine residues (H27, H40 and H92), four phenylalanine residues (F48, F50, F80 and F100), nine tyrosine residues (Y4, Y11, Y24, Y38, Y42, Y45, Y56, Y57 and Y68) and one tryptophan residue (W59). Upon inhibitor binding, some of the aromatic proton and the high field shifted methyl proton resonances of RNase T_1 were significantly shifted, suggesting that these residues are involved in the binding of the inhibitors. To elucidate the interaction of RNase T_1 with inhibitors, the assignments of the aromatic proton and the high field shifted methyl proton resonances of RNase T_1 and its inhibitor complexes are essential. The assignment of these proton resonances of RNase T_1 in the 2'GMP bound form was made using DQF-COSY, HOHAHA and NOESY in the presence of 2'GMP at a molar ratio of 3.0. The sequence-specific resonance assignment of those aromatic proton and high field shifted methyl proton resonances was subsequently made as compared to the results of the NOESY spectrum and the molecular structure of the RNase T_1-2'GMP complex elucidated by the X-ray crystallography. Based on the results, the aromatic proton and the high field shifted methyl proton resonances

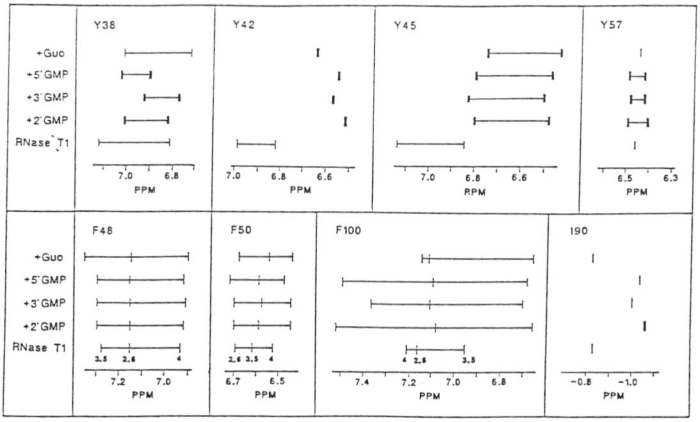

Fig. 1 Comparison of the chemical shifts of the aromatic and the high field shifted methyl proton resonances of RNase T_1.

of RNase T_1-3'GMP, RNase T_1-5'GMP and RNase T_1-Guo were also assigned to the specific amino acid residues.

The aromatic proton resonances were classified into several groups according to the behaviour of the chemical shift upon inhibitor binding. Y42 and Y45 showed slow exchange process compared to the chemical shift time scale for all inhibitors studied in the present study. Y38 and Y57 also showed slow exchange process in case of 2'GMP, 3'GMP and 5'GMP binding but not for Guo binding. F100 showed large change of chemical shifts upon binding of the inhibitors with fast exchange process, while other aromatic proton resonances (Y4, Y11, Y24, Y56, F48, F50, F80 and W59) did not show any appreciable chemical shift changes. In Figure 1, we compare the chemical shifts of some of the aromatic proton and I90 δ methyl proton resonances of the RNase T_1-inhibitor complexes with reference to the chemical shifts of those of free RNase T_1. It is to be noted that the induced chemical shifts of the Y42, Y45, V79, A87, Y38, Y57, F100 and I90 proton resonances upon binding of 5'GMP are quite similar to those upon binding of 2'GMP and 3'GMP (Figure 1), suggesting that the guanine bases and the phosphate groups of these inhibitors bind to the the similar sites on the enzyme. This supports the that there is a conformational difference among those inhibitors around the glycosidic linkages (syn; 2' and 3'GMP, anti; 5'GMP). The present result suggests that RNase T_1 has no intrinsic binding site for the phosphate moiety at 5' position. The conformational change around glycosidic bond in 5'GMP is possible only when the sugar moiety does not form the hydrogen bonds with RNase T_1.

Acknowledgments

We are grateful to Prof. Ken-ichi Tomita, Dr. Toshio Hakoshima and Dr. Satoshi Nishikawa of Osaka University for their helpful discussions.

Symposium Series No. 21 1989 Nucleic Acids Research

Stacking interactions between protonated nucleic acid bases and aromatic amino acids: spectroscopic and structural analyses of m^3CMP-tryptophan derivative complex

H.Ueda, K.Segawa, M.Doi, M.Inoue, and T.Ishida

Osaka University of Pharmaceutical Sciences, 2-10-65 Kawai, Matubara, Osaka 580, Japan

ABSTRACT

As a stacking model between nucleic acid bases and aromatic amino acids, the interaction on m^3CMP-tryptophan derivative has been studied by ^{1}H-NMR and X-ray crystal analyses. From the comparative ^{1}H-NMR experiments using CMP and m^3CMP, it is suggested that the N(3)-protonation by methylation greatly strengthens the stacking interaction with tryptophan. Parallel alignment with a separation distance of 3.38A is shown by the X-ray analysis of m CMP-tryptamine complex. The stacking mode is very similar to those observed in the complexes of indole ring with m^1A and m^7G.

INTRODUCTION

The methylation of nucleic acis bases is very significant for biological process, as can be seen in the cap terminal structure of eukaryotic mRNA. As a series of elucidating the biological implication of N-quarternization caused by the base methylation, recently we reported the significant stacking interactions of aromatic amino acids with m^1A and m^7G derivatives[1)-3)]. Here we report the stacking mode of m^3C-tryptophan complex studied by ^{1}H-NMR and X-ray crystal analyses.

RESULTS and DISCUSSION

In order to examine the structural stability accompanying the complex formation,the association constant and thermodynamic parameters(ΔH andΔS) were evaluated by the ^{1}H-NMR experiments.

These data were determined from the temperature dependence(30 °C-70 °C) of the upfield shift changes of H1'(CMP and m^3CMP) and methyl(m^3CMP) protons as a function of L-tryptophan methylester (TrpOMe) concentration (25-300mM). The results show that the indole ring of TrpOMe interacts with m^3C base about four times more than with C base. Furthermore, -ΔH and -ΔS values, which are closely related to the magnitude of the force of ring stacking formation and the order of the stacking structure, respectively, suggest that the complex structure of m^7GMP-TrpOMe is thermodynamically stable. These insight clearly prove that the protonated nucleic acid base has a large affinity for the aromatic amino acid such as tryptophan, compared with the neutral base.

X-ray analysis was carried out for the m^3CMP-tryptamine complex crystallized from 20% aqueous methanol solution. The interaction mode observed is shown in Fig.1. The dihedral angle between the m^3C base and indole ring is 2.7°(2), and the mean separation distance is 3.38A. This stacking characteristic with indole ring has also been observed in the complex crystals of m^7G and m^1A bases[1),2)], while such significant interaction has not sofar been reported at the crystalline state.

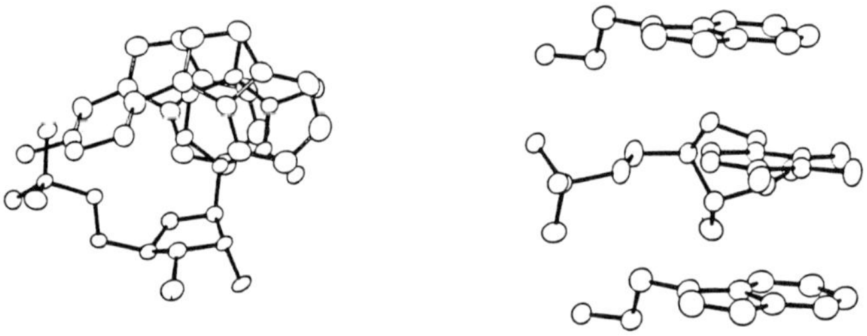

Fig.1 Stacking Mode Observed in m^3CMP-Tryptamine Complex Crystal

REFERENCES

1. Ishida,T.,Doi,M.,Ueda,H.,Inoue,M. and Scheldrick,G.,M. (1988) J.Am.Chem.Soc. 110,2286-2294.
2. Isida,T.,Shibata,M.,Fujii,K. and Inoue,M. (1983) Biochemistry 22,3571-3581
3. Kamiichi,K.,Doi,M.,Nabae,M.,Ishida,T. and Inoue,M. (1987) J.Chem.Soc.Perkin Tran.2 1739-1745.

Conformational characteristics of 4-acetylcytidine found in tRNA

Gota Kawai, Takeshi Hashizume[1], Tatsuo Miyazawa, James A.McCloskey[1], and Shigeyuki Yokoyama[2]

Faculty of Engineering, Yokohama National University, Tokiwadai, Hodogaya-ku, Yokohama 240, Japan, [1]Departments of Medicinal Chemistry and Biochemistry, University of Utah, Salt Lake City, Utah 84112, USA, and [2]Department of Biophysics and Biochemistry, Faculty of Science, University of Tokyo, Hongo, Bunkyo-ku, Tokyo 113, Japan

ABSTRACT

The conformational characteristics of a modified nucleoside 4-acetylcytidine (ac^4C) was analyzed by proton nuclear magnetic resonance spectroscopy. This nucleoside was found to take the *C3'-endo* form predominantly. The enthalpy difference between the *C2'-endo* form and the *C3'-endo* form of the ribose moiety of ac^4C was determined and 4-acetylation was found to stabilize the *C3'-endo* form further by 0.83 $kcal \cdot mol^{-1}$. This conformational characteristics of ac^4C is probably important for the correct codon recognition and the stability of the tertiary structure of tRNA molecule.

INTRODUCTION

4-Acetylcytidine (ac^4C) has been found in the first position of the anticodon of the elongator methionine tRNA from *Escherichia coli* and the glutamine, glutamate, lysine, and proline tRNAs from halobacteria. Such an ac^4C residue is probably important for the correct codon recognition. Further, ac^4C has also been found in position 12 of the D-stem of eukaryotic leucine and serine tRNAs. In order to elucidate the role of ac^4C, in the present study, the conformational characteristics of ac^4C and cytidine (C) were analyzed by the proton nuclear magnetic resonance (^1H-NMR) spectroscopy.

RESULTS AND DISCUSSION

3 mg of ac^4C or C was dissolved in ^2H_2O and 400-MHz ^1H-NMR spectra were recorded on a Bruker AM-400 spectrometer. The fractional populations of the *C3'-endo* form and the *C2'-endo* form at 25°C were determined as 74 and 26 %, respectively, for ac^4C and 60 and 40 %, respectively, for the unmodified C. This indicates that the *C3'-endo* form is further stabilized in ac^4C

as compared with C.

From the temperature dependence of fractional populations, the enthalpy difference between the *C3'-endo* form and the *C2'-endo* form was determined as 1.22 and 0.39 kcal·mol^{-1} for ac^{4}C and C, respectively. Thus 4-acetylation of C was found to stabilize the *C3'-endo* form by 0.83 kcal·mol^{-1}. This stabilization of the *C3'-endo* form is even stronger than that by 2-thiolation of uridine (0.75 kcal·mol^{-1}) (1). 4-Acetylation will effectively stabilize the A-type conformation (with the *C3'-endo* form of the ribose ring) of RNA and contribute to the correct recognition of codons and to the maintenance of active conformations of tRNA molecules as is the case for 2-thiolation (1).

The ac^{4}C residue in the first position of the anticodon of the elongator methionine tRNA from *Escherichia coli* is to recognize only G as the third letter of codon. For the recognition of such a G residue of codon, the C residue in the first position of the anticodon is to take the A-type conformation. The stablization of the A-type conformation by 4-acetylation of C will contribute to the correct codon recognition. Note that ac^{4}C in position 12 of the D-stem has been exlusively found in eukaryotic tRNAs specific for leucine and serine; possibly this residue is recognized by leucyl and seryl-tRNA synthetases of eukaryote.

In the present study, 4-acetylation of C was found to significantly affect the conformational characteristics, as were the cases for 2-thiolation and 5-substitution of uridine and for 2'-*O*-methylation of pyrimidine nucleotides. Through such regulation of conformational properties, all these modifications are important for the biological functions of tRNA species.

The molecular mechanism in the significant stabilization of the *C3'-endo* form of C by 4-acetylation is yet to be elucidated. Therefore, more detailed conformational analyses (including nuclear Overhauser effect) of ac^{4}C and related nucleotides are now in progress in our laboratories.

REFERENCES

1. Yokoyama, S., Watanabe, T., Murao,K., Ishikura,H., Yamaizumi, Z., Nishimura, S. and Miyazawa, T. (1985) *Proc. Natl. Acad. Sci. USA* **82**, 4905-4909.

Induced circular dichroism in nucleic acid–acridine derivative complexes

S.Tani and Y.Kubota

Department of Chemistry, Faculty of Science, Yamaguchi University, Yoshida, Yamaguchi 753, Japan

ABSTRACT

Visible absorption and circular dichroism (CD) spectra have been measured for complexes formed between nucleic acids (calf thymus DNA, poly(rA)·poly(rU) and poly(rI)·poly(rC)) and 9-aminoacridines (quinacrine, acranil and 9-amino-6-chloro-2-methoxy acridine). With poly(rA)·poly(rU), a new absorption band was observed at longer wavelengths. The nucleic acid-drug complexes showed considerable different induced CD spectra. Analysis of these CD spectra suggests that the cationic side chains of quinacrine and acranil play an important role on the binding properties to DNA and poly(rA)·poly(rU).

INTRODUCTION

The interaction of the antimalarial drug quinacrine (QAC) with nucleic acids has been extensively investigated using spectrophotometric, fluorescence and NMR techniques[1-3]. Quinacrine exists as a dication in neutral aqueous solution and the binding of the drug to DNA can occur first by intercalation of the acridine ring between DNA base pairs and second by electrostatic binding of the protonated side chain, at the 9-position in the aromatic acridine ring, to the DNA phosphate groups. CD spectroscopy is one of the most useful techniques in monitoring the interaction and characterizing the conformation of nucleic acid-drug complexes. In order to clarify the influence of the nucleic acid conformations and of the side chains of quinacrine and acranil on the binding interactions, we have investigated induced CD spectra of quinacrine, acranil (ACR) and 9-amino-6-chloro-2-methoxy acridine (ACMA) complexed with native and synthetic nucleic acids (calf thymus DNA, poly(rA)·poly(rU) and poly(rI)·poly(rC)).

RESULTS

Visible absorption and CD spectra of various nucleic acid-drug complexes were measured as a function of P/D ratio (DNA-phosphate to drug concentrations) with a Hitachi U-3200 spectrophotometer at 25 °C and a Jasco J-20 spectropolarimeter at room temperature, respectively, in 5 mM phosphate buffer (ionic strength of 0.05, pH 7). The results are summarized in Table I.

A new absorption band in the wavelength near 460-475 nm is observed as a shoulder for poly(rA)· poly(rU)-drug complexes at high P/D ratios in the 0.05 ionic strength buffer; the shape of the new band does not depend on the ionic strength. Since poly(rA)· poly(rU) and poly(rI)· poly(rC) belong to the A-family and their conformations are very similar, the phenomenon mentioned above is attributable not to slight different conformations of these polymers but to the unique interaction of the aromatic ring of drugs with successive adenine-uracil base pairs.

The CD spectra of the DNA-ACR and -QAC complexes are very complicated in the wavelength range 375-500 nm. This observation shows that acranil, like quinacrine, can intercalate between DNA base pairs. The DNA-drug interactions are also characterized by the appearance of isodichroic points in high P/D ranges. The CD spectra of poly(rA)· poly (rU)-ACMA and -ACR complexes are very simplified, having only slight optical activity in the above wavelength range, and very similar to each other. However, it is found that poly(rA)· poly(rU)-QAC complex has a simple and negative CD band in the 400-480 nm range. The CD spectra obtained for poly(rI)· poly(rC)-drug complexes are very similar to the visible absorption spectra of free drugs themselves. Analysis of the CD spectra shows that the cationic side chains significantly affect the CD spectra and the electrostatic interaction between the side chains and the phosphate groups of DNA and poly(rA)· poly(rU) play an important role on the binding. Such a remarkable influence of the side chains is not found on the complexation with poly(rI)· poly(rC).

The authors thank Professors M. Fujimoto and T. Oishi for their supports in CD measurements.

Table I. Visible absorption and CD spectra of nucleic acid - dye complexes

Complex	P/D	visible absorption λ/nm ($\varepsilon \times 10^{-3}$/$M^{-1}cm^{-1}$)	CD[a] λ/nm ($\Delta\varepsilon$/$M^{-1}cm^{-1}$)
Calf thymus DNA-QAC	100	455(5.60), 433(5.58) 349(4.06)	475(+0.35), 456(-0.56), 429(-0.27) 335(+1.34)
Calf thymus DNA-ACR	20	454(4.67), 433(4.70) 351(3.55)	472(+0.46), 455(-0.52), 429(-0.38) 340(+2.08)
Calf thymus DNA-ACMA	51	441(3.84),419(3.97) 344(3.67)	435(+0.79), 415(+0.89), 343(+1.53)
Poly(rA)· poly(rU)-QAC	101	451(5.53), 348(3.93)	440(-0.43), 348(+1.91), 333(+1.67)
Poly(rA)· poly(rU)-ACR	41	451(4.63), 352(3.86)	351(+3.33)
Poly(rA)· poly(rU)-ACMA	45	442(3.74), 418(3.84) 344(4.39), 330(2.93)	442(n), 400(p), 344(+1.66), 330(+1.59)
Poly(rI)· poly(rC)-QAC	49	453(5.73), 349(4.03)	450(+0.74), 348(+2.55)
Poly(rI)· poly(rC)-ACR	28	453(4.92), 351(4.14)	448(+0.93), 349(+3.16)
Poly(rI)· poly(rC)-ACMA	30	441(3.25), 418(4.40) 344(2.62)	438(+0.57), 410(+0.69), 344(+0.86)

[a] n = negative, p = positive.

REFERENCES

1 Krey, A. K. & Hahn, F. E. (1974) Mol. Pharmacol. **10**, 686-695.
2 Weisblum, B. & de Haseth, P. L. (1972) Proc. Natl. Acad. Sci. USA **69**, 629-632.
3 Jones, R. L. & Wilson, W. D. (1980) J. Am. Chem. Soc. **102**, 7776-7778.

NMR studies on DNA hairpin structures with a five nucleotide loop

Seiichi Uesugi, Yasushi Oda, Hidekazu Hiroaki, and Morio Ikehara[1]

Faculty of Pharmaceutical Sciences, Osaka University, and [1]Protein Engineering Research Institute, Suita, Osaka 565, Japan

ABSTRACT

One- and two-dimensional NMR experiments have been undertaken to investigate the structure of DNA hairpins with a five nucleotide loop. Analysis of proton NMR spectra suggests that the four hairpin structures examined have some common structural features; B-type conformation in the stem region and the same stacking pattern, 5'(XXX-turn-XX)3', in the loop region. The phosphorus NMR spectra suggest that the conformational changes in the loop region affect the backbone conformation of the stem duplex.

INTRODUCTION

Hairpins are one of the secondary structures for nucleic acids and contain a single-stranded loop region and a double-stranded stem region. We have studied conformation of four related DNA hairpins which have five unpaired nucleotides in the loop region by one- and two-dimensional NMR spectroscopy.

	1 2 3 4 5 6 7 8 9 10111213
(hairpin I)	d-G G C G-T T T T T-C G C C
(hairpin II)	d-G G C G-A A A A A-C G C C
(hairpin III)	d-G G C G-A A T A A-C G C C
(hairpin IV)	d-G G C G-A A A T A-C G C C

MATERIALS AND METHODS

The oligonucleotides were synthesized by the phosphotriester method using dimer blocks. NMR experiments were performed on a JEOL GX500 spectrometer. MNR samples were prepared in H_2O or D_2O (0.4 ml) containing 0.1 M NaCl, 10 mM sodium phosphate buffer(pH 7.1).

RESULTS AND DISCUSSION

Melting temperatures for the oligomers I∿IV were not dependent on oligomer concentrations indicating formation of a hairpin. In hairpin II, sequential NOE connectivities between H8/H6 of the residue (n) and H1'(or H2'/H2") of the residue (n-1) were observed through the sequence except for that between A8 and A7. The pattern of inter-residual NOE connectivities suggests that hairpin II has a B-type duplex in the stem and regularly stacked bases in the loop, and a turn in the loop occurs between residues A7 and A8. H4' and H5'/H5" protons of the residue A7 in hairpin II showed unusual upfield shifts (∿1 ppm), but not in hairpin I. This upfield shift may result from a shielding effect of the adenine base of A8 residue located at the turn. This explanation was confirmed by measuring ^{1}H NMR spectra of hairpins III and IV. As expected, the upfield shift was observed only in hairpin III, which has A at the 8th position. Above results suggest that these hairpins with a five nucleotide loop have the same stacking pattern in the loop.

In order to examine the sugar-phosphate backbone conformation, ^{31}P NMR spectra were measured. In hairpin I, one of the phosphorus signals, assigned to C10pG11, shows a large downfield shift (∿-2.5 ppm) indicating an altered phosphate conformation. Such an unusual shift was not observed in the other hairpins. We suppose that the loop in hairpin I takes a unique conformation as a result of inter-residual interaction between the loop T residues, and this conformational change in the loop affects the stem conformation.

The molecular structure of Okazaki fragment r(CGCA)d(AAAAAGCG):d(CGCTTTTTGCG) in solution

Satoshi Fujii, Sachiko Matsui, Ken-ichi Tomita, Masaya Orita, Toshiki Tanaka, and Seiichi Uesugi

Faculty of Pharmaceutical Sciences, Osaka University, 1–6 Yamadaoka, Suita, Osaka 565, Japan

ABSTRACT

The hetero duplex molecule, r(CGCA)d(AAAAAGCG):d(CGCTTTTTG CG) which corresponds to Okazaki fragment was synthesized and its molecular structure has been analyzed by NMR study. The RNA strand of RNA-DNA hybrid region adopts A-form and DNA strand of the same region deviates from the standard B-form. The conformation of DNA-DNA duplex segment belongs to B-form. The hybrid-DNA duplex junction shows a structural discontinuities, A-B junction. The same conformational characteristic of oligo(dA): oligo(dT) tract as that of DNA oligomer which has same base sequence has been observed.

INTRODUCTION

In DNA replication, Okazaki fragments are initiated by a short primer containing a few ribonucleotides. The DNA-RNA hybrid and its attached DNA duplex may serve as a model for visualizing the structure of the Okazaki DNA fragment after initiation of replication. We have undertaken an NMR study of title compound. The DNA oligomer (noted as DODE1) which has the same sequence had been also studied.[1] There is a oligo(dA): oligo(dT) tract which is the cause of DNA bending in the middle part of molecule. The numbering of the residues is as follows.

```
        1  2  3  4    5  6  7  8  9 10 11 12
5'    r(C -G -C -A)-d(A -A -A -A -A -G -C -G)  3'
3'    d(G -C -G -T    -T -T -T -T -T -C -G -C)  5'
       24 23 22 21   20 19 18 17 16 15 14 13
```

METHOD

Both oligonucleotides were synthesized by the solid-phase phosphoramidite method[2]. The NMR spectra were recorded on a JEOL GX-500. The theoretical chemical shifts were calculated using the computer program called SHIFTS[3] modified by S.F.

RESULT & DISCUSSION

Base protons and H1's were assigned in a sequential manner and by referring the results of DODE1. The theoretical polymerization effect of chemical shifts (PARA1) were calculated for several models. By using this PARA1 and the observed movement of chemical shift from one of DODE1 (PARA2), the general conformations for various region can be estimated. The observed chemical shifts of DNA-DNA duplex region are not different from those of DODE1 which adopts B-form family structure. On the other hand, the some protons of DNA-RNA hybrid region are different beyond 0.15ppm from DODE1. PARA2 shows the A-type geometry for RNA strand of hybrid region and the highest correlation coefficient of PARA1 is also A-type. In the case of DNA strand of hybrid region, A-type gives good coefficient but the PARA2 couldn't estimate as exactly A-form but non standard B-from. A-type like PARA2's are observed among A5 base protons, suggesting the presence of A-type base stacking mode in the junction region.

As same as DODE1, the interesting pattern of the cross peaks between the H2 of the adenine and the H1' of the 3' neighbor on the same strand and the 3' neighbor of the complementary residue is observed. The strong intensities of interstrand cross peaks shows that the width of the minor groove of oligo(dA):oligo(dT) tract becomes narrower from the classical B-form.

REFERENCES

1) Katahira, M., Sugeta, H., Kyogoku Y., Fujii, S., Fujisawa, R. and Tomita, K.(1988) Nucleic Acids Res., 16, 8619-8632.
2) Usman, N.,Ogilvie, K.K., Jiang, M.-Y. and Cedevguen, R.J. (1987) J. Am. Chem. Soc., 109, 7845.
3) Giessner-Prettre, C., Prado, F.R., Pullman, B., Kan, L.-S., Kast, J.R. and Ts'o, P.O.P. (1981) Computer Prog. Biomed., 13, 167-184.

A new model for DNA bending on the basis of gradual compression of the minor groove from 5′ to 3′ along the oligo(dA) tract

Masato Katahira, Hiromu Sugeta, and Yoshimasa Kyogoku

Institute for Protein Research, Osaka University, Suita, Osaka 565, Japan

ABSTRACT

The conformations of double stranded d(GGAAATTTCC)x2, d(GGTTTAAACC)x2, d(CGCAAAAAAGCG)·d(CGCTTTTTTGCG) and d(GCATTTTGAAACG)·d(CGTTTCAAAATGC) have been studied by means of NMR spectroscopy. Analyses of cross peaks in NOESY spectra between H2 of an adenine and H1' of the deoxyribose in the 3'-neighbouring residue on the complementary strand revealed that the minor groove of the oligo(dA) tract is compressed gradually from 5' to 3' along the tract in four oligonucleotides. A new model is proposed as to DNA bending based on the evidence of gradual compression of the minor groove. This model can explain why $d(GGAAATTTCC)_n$x2 and $d(GAAAATTTTC)_n$x2 are bent, and $d(GGTTTAAACC)_n$x2 and $d(GTTTTAAAAC)_n$x2 are not bent. The bending of $d(A_jN_{10-j})_n$x2 sequences can also be explained.

INTRODUCTION

It has been demonstrated that DNAs which contain the oligo(dA) tract are bent by means of electrophoresis and electron microscopy. Several models for DNA bending have been proposed [1-4], but they are contradictory. We have already studied the conformation of d(CGCAAAAAAGCG)·d(CGCTTTTTTGCG) by means of NMR spectroscopy [5]. To further clarify the correlation between the oligo(dA) tract and DNA bending, we have performed studies of the conformations of four oligonucleotides.

MATERIALS AND METHOD

NMR data were recorded on a JEOL GX-500. One dimensional and two dimensional NOESY spectra in H_2O were obtained by 1-1 or 1-1 echo pulses. NOESY spectra in D_2O were recorded with several short mixing times (60, 90, 100, 120 and 150 ms). Volume integration of each cross peak was carried out and interproton distances were obtained.

RESULTS AND DISCUSSION

It is confirmed that the minor groove of the oligo(dA) tract is compressed gradually from 5' to 3' along the tract by the analysis of a distance between H2 of an adenine and H1' of the deoxyribose in the 3'-neighbouring residue on the complementary strand. Compression of the minor groove leads to bending toward the minor groove. The H2-H1' distance in a classical B-DNA model is 5.1 Å. Difference between 5.1 Å and an experimentally obtained distance is regarded as a yard stick of an extent of DNA bending. To take into account the direction of bending in a helix, each residue of the oligonucleotide is arranged on a circle with a 36 ° interval. Thus an extent and direction of bending can be expressed by the vector at each residue and a sum of each vector indicates total bending.

A sum of vectors in the case of d(GGAAATTTCC)x2 is greater than that in the case of d(GGTTTAAACC)x2. In the latter case, the two longest vectors point in opposite directions and thus cancel each other out, although in the former case these two vectors point in similar directions, and thus are most effectively additive. Thus it can be understood why $d(GGAAATTTCC)_nx2$ is bent and $d(GGTTTAAACC)_nx2$ is not so bent, if any at all. In a similar way, difference of bending between $d(GAAAATTTTC)_nx2$ and $d(GTTTTAAAAC)_nx2$ can be explained.

Moreover, the extent of bending of $d(A_jN_{10-j})_nx2$ sequences can also be explained generally in view of the gradual compression of the minor groove along the oligo(dA) tract.

REFERENCES

1. Koo,H.S., Wu,H.M. and Crothers,D.M. (1986) Nature, 320, 501-506.
2. Ulanovsky,L.E. and Trifonov,E.N. (1987) Nature, 326, 720-722.
3. Burkhoff,A.M. and Tullius,T.D. (1987) Cell, 48, 935-943.
4. Calladine,C.R., Drew,H.R. and McCall,M.J. (1988) J. Mol. Biol., 201, 127-137.
5. Katahira,M., Sugeta,H., Kyogoku,Y., Fujii,S., Fujisawa,R. and Tomita,K. (1988) Nucleic Acids Res., 16, 8619-8632.

Three-dimensional structure of ribonuclease Ms*3′-guanylic acid complex at 2.5 Å

Takamasa Nonaka, Yukio Mitsui, Hideaki Watanabe[1], Kazuko Ohgi[1], Masachika Irie[1], Hiroshi Mizuno[2], and Kazuo T.Nakamura[3]

Faculty of Engineering, Nagaoka University of Technology, Nagaoka, Niigata 940-21, [1]Department of Microbiology, Hoshi College of Pharmacy, Shinagawa-ku, Tokyo 141, [2]National Institute of Agrobiological Resources, Tsukuba, Ibaragi 305, and [3]Faculty of Pharmaceutical Sciences, University of Tokyo, Bunkyo-ku, Tokyo 113, Japan

Ribonuclease Ms (RNase Ms; EC 3. 1. 4. 23) was isolated and purified from *Aspergillus saitoi* by Ohgi and Irie[1]. RNase Ms hydrolyzes RNA completely to mononucleotides *via* 2′,3′-cyclic nucleotides without absolute base specificity. The enzyme consists of a single polypeptide chain (Mr 11,401) of 105 amino acid residues, the sequence of which was determined by Watanabe *et al.* It has two disulfide bonds (Cys3-Cys11, Cys7-Cys102) similar to RNase T_1 (2-10, 6-103). There are many regions of very similar sequences between RNase Ms and RNase T_1 despite large differences in base specificity. The two enzymes have identical amino acids for 65% of their aligned sequences. In order to determine the three-dimensional structure of RNase Ms and elucidate the origin of its lack of base specificity, we solved the crystal structure of RNase Ms*3′-guanylic acid (3′-GMP) complex using molecular replacement methods based on the atomic co-ordinates of RNase T_1.

The complex was crystallized using the vapour diffusion technique in the hanging-drop mode. Large prism crystals were obtained in hanging-drops containing 1.7mM RNase Ms, 8.0mM 3′-GMP, 27% (v/v) 2-methyl-2,4-pentanediol and 35 mM acetate (pH 6.0). The crystals belong to an orthorhombic space group $P2_12_12_1$ with cell dimensions a = 47.0 Å, b = 62.8 Å, c = 37.9 Å. A total of 4195 independent reflections in the resolution range of 20-2.5 Å were collected on a Rigaku AFC-5R four-circle diffractometer.

The search molecule in a molecular replacement study was constructed as follows: Amino acid residues of RNase T_1 identical with those of RNase Ms were preserved. Homologous and non-homologous residues were replaced with Ala and Gly, respectively. The search molecule was put in a triclinic *P1* cell with dimensions of 80x80x80 Å. The fast rotation function was calculated to determine the orientation of the search molecule in the unit cell. The highest peak was 5.1σ higher than the mean and 1.4σ higher than the second highest. Then a translational search was carried out to locate the search molecule in the unit cell. The highest peak was 7.6σ higher than the mean and 1.6σ higher than the second highest. The refined Eulerian angles and translation vector of the search molecule are (α=124.7°, β=34.1°, γ=61.4°) and (15.7Å, 16.8Å, 2.1Å), respectively.

The crystal structure has been refined to an R-factor of 0.231 by a stereochemically restrained least-squares method using diffraction data in the resolution range of 5.0-2.5 Å. The r.m.s. deviation of bond lengths from ideal values is 0.024 Å. The starting model of the refinement was constructed by deleting 2 residues of the search molecule, replacing 13 residues with the corresponding ones of RNase Ms and inserting 3 residues of RNase Ms to the search molecule. Both difference Fourier and fragment-deleted Fourier maps were used in the model manipulating process. Stereoscopic drawing of RNase Ms•3′-GMP complex is shown below. The overall three-dimensional structure of RNase Ms is similar to that of RNase T_1. With regard to 3′-GMP molecule, the electron density for guanine base and ribose is well defined, while there is only low electron density for the 3′-phosphate group. Further refinement is in progress.

REFERENCES

1 Ohgi, K. & Irie, M. (1975) *J. Biochem. (Tokyo)* 77, 1085-1094.

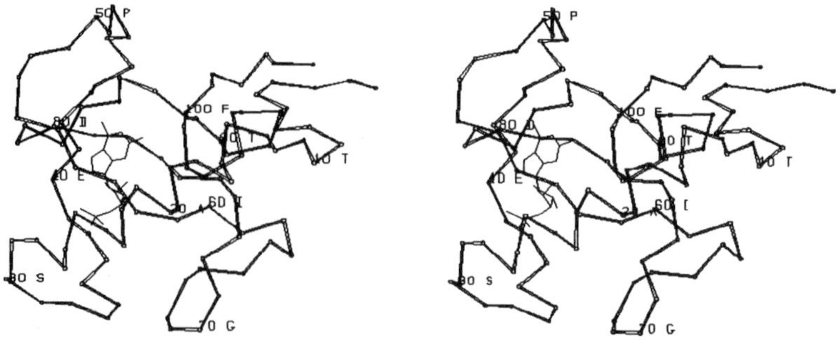

New fluorinated carbocyclic nucleosides: synthesis and bioactivity

Toshiaki Nakayama, Yoshitomi Morizawa, Yasushi Matsumura, Arata Yasuda, and Keiichi Uchida

Research Center, Asahi Grass Co., Ltd., Hazawa, Kanagawa-ku, Yokohama 221, Japan

ABSTRACT

Novel fluorinated carbocyclic nucleosides, (±)-3'-deoxy-3'-fluoroaristeromycin and (±)-2',3'-dideoxy-3'-fluoroaristeromycin have been synthesized. The antitumor activities of these compounds were studied.

INTRODUCTION

Fluorinated nucleosides, represented by 3'-deoxy-3'-fluoroadenosine[1] and 2',3'-dideoxy-3'-fluorothymidine[2], show the remarkable biological activities such as antitumor and antiviral ones. These activities are thought to be brought about by the mimic effect of fluorine atom attached on 3'-position, the enzymatically important site. We have been attracted by nucleosides bearing fluorine atom on C-3' and planed to make more stable carbocyclic analogues.

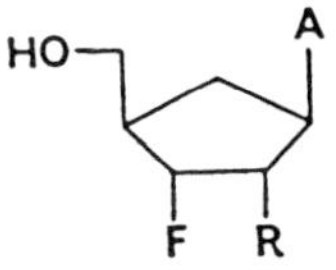

R : OH **1**
H **2**

We now report the stereoselective synthesis and in vitro antitumor activities of (±)-3'-deoxy-3'-fluoroaristeromycin **1** and (±)-2,'3'-dideoxy-3'-fluoroaristeromycin **2**.

RESULTS

Synthesis of(±)-3'-deoxy-3'-fluoroaristeromycin 1

The synthetic route of compound **1** is outlined in Scheme. The epoxide **3**, derived from 2-cyclopentenemethanol, was treated with NaN_3 to give azide alcohol. Protection of free hydroxyl group and deprotection of $^tBuMe_2Si$ group afforded compound **4a** (R=H). The key step is a S_N2 displacement of OR group with fluoride ion. After several trials, we successfully treated **4b** (R=SiMe_3) with piperidinosulfur trifluoride followed by the hydrogenation of benzyl group and azide group to give the desired carbocyclic sugar with cis-fluorohydrin moiety **5**. The adenyl substitutent was elaborated by a known sequence[3].

Synthesis of (±)-2',3'-dideoxy-3'-fluoroaristeromycin 2

Cis-4-acetamido-1-acetoxymethyl-2-cyclopentene **6** was treated with HF-pyridine complex and N-bromosuccinimide[4] to afford trans fluoro-bromide **7** in 77% yield (regioselectivity: >90%). nBu_3SnH reduction of compound **7** followed by deprotection gave **8** in 85% yield. Fluoropentylamine **8** was transformed to 2',3'-dideoxy-3'-fluoroaristeromycin **2** according to the same procedure as **1**.

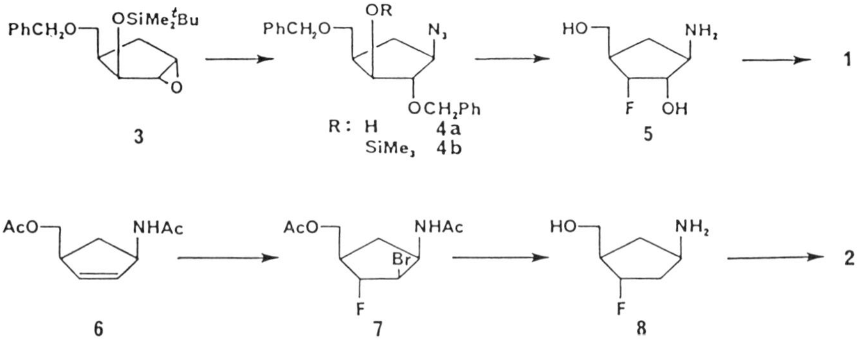

Antitumor activity

(±)-3'-deoxy-3'-fluoroaristeromycin **1** and its deoxy analogue **2** were tested for the ability to inhibit the growth of L1210 leukemia in vitro. The IC_{50} value are as follows: **1** : >1000 μg/ml, **2** : 3 μg/ml.

REFERENCES

1. Y. Morizawa, T. Nakayama, A. Yasuda, K. Uchida, and T. Sasaki, JCU Pharm Sci 1987 poster abstracts, 223.
2. G. Etzold, R. Hintsche, G. Kowollik, and P. Langen, Tetrahedron, **29**, 2463 (1971).
3. Y. F. Shealy, and J. D. Clayton, J. Am. Chem. Soc., **88**, 3885 (1966), ibid., idem, **91**, 3075 (1969).
4. G. A. Olah, M. Nojima, and I. Kerebes, Synthesis, 780 (1973)

Novel synthesis of carbocyclic nucleosides from 2-azabicyclo[2.2.1]heptanes via the reductive N_2-C_3 cleavage reaction as a key step

Nobuya Katagiri*, Makoto Muto, and Chikara Kaneko

Pharmaceutical Institute, Tohoku University, Aobayama, Sendai 980, Japan

ABSTRACT

Carbocyclic ribofuranosyl and arabinofuranosyl pyrimidine nucleosides were synthesized stereoselectively from 2-azabicyclo[2.2.1]hept-5-en-3-one via $NaBH_4$ promoted C-N bond cleavage reaction as a key step.

INTRODUCTION

Previously, we reported the stereo- and enantio-selective syntheses of C-nucleosides and their carbocyclic analogues from Diels-Alder adducts using reductive retrograde aldol (RRA) reaction (K_2CO_3-$NaBH_4$/MeOH) as a key step.[1-3] This paper deals with the synthesis of carbocyclic pyrimidine nucleosides from 2-azabicyclo[2.2.1]heptan-3-ones[4], essentially by the same methodology.

RESULTS

2-Azabicyclo[2.2.1]hept-5-en-3-one (**1**), prepared by hetero Diels-Alder reaction of tosyl cyanide with cyclopentadiene,[4] reacted with chloroacetyl isocyanate at room temperature to give the N-carbamoyl bicyclic compound, whose oxidation with OsO_4 followed by acetonization gave the acetonide **2a**. Reductive C-N bond cleavage of **2a** with $NaBH_4$ in methanol at room temperature was accompanied by removal of the chloroacetyl group to give the urea derivative **3** (R=H). Usual construction of pyrimidine ring from urea derivatives gave rise to desired carbocyclic pyrimidine nucleosides **4**.

A more direct approach for the synthesis of **4** was also accomplished. Thus, reaction of **1** with 3-alkoxy-acryloyl(or -

methacryloyl)isocyanate followed by the same manipulation as above gave the acetonide **2b.** **2b** was then subjected to the C-N bond cleavage reaction to give the N-acylurea **3b.** Intramolecular cyclization of **3b** under acidic conditions produced **4** (Y=H or Me, Z=OH).

The same methodology has also been applied to the synthesis of carbocyclic arabinofuranosyl nucleosides **8.** The epoxide **5** obtained from **1** by m-CPBA, when reacted with various isocyanates, gave the N-carbamoyl compounds **6.** By treatment with $NaBH_4$, the oxazolidine derivatives **7**, expected as synthons for **8**, was obtained in almost quantitative yield. The direct formation of **7** is best explained by assuming that the C-N bond fission brings about concomitant epoxide ring cleavage.

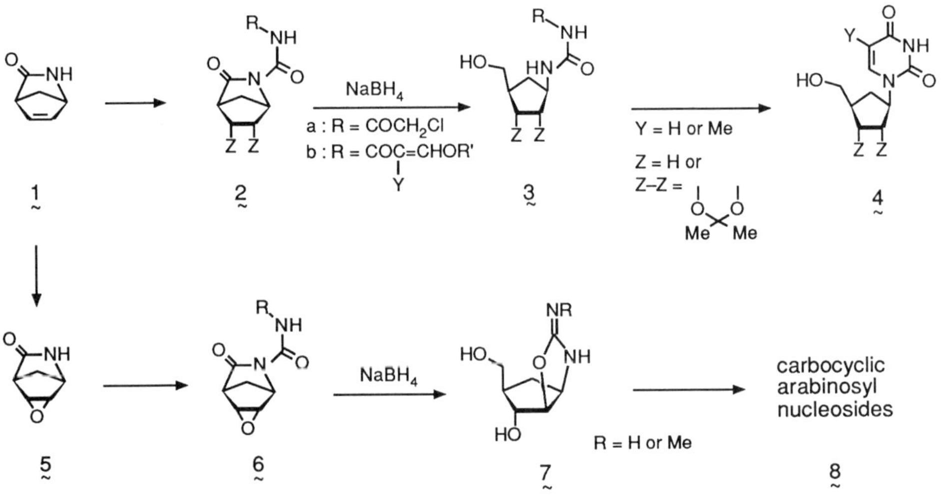

REFERENCES AND NOTES

1. Katagiri, N., Haneda, T., Hayasaka, E., Watanabe, N., and Kaneko, C. (1988) J. Org. Chem., **53**, 226-227.
2. Katagiri, N., Akatsuka, H., Kaneko, C., and Sera, A. (1988) Tetrahedron Lett., **29**, 5397-5400.
3. Katagiri, N., Akatsuka, H., Haneda, T., Kaneko, C., and Sera, A. (1989) J. Org. Chem., **53**, 5464-5470.
4. A part of this work was reported as a communication: Katagiri, N., Muto, M., and Kaneko, C.(1989) Tetrahedron Lett., **30**, 1645-1648 and references cited therein.

Organoselenium-mediated addition of carbon radicals to 4′,5′-unsaturated uracil-nucleosides and its stereocontrol

Kazuhiro Haraguchi, Hiromichi Tanaka, and Tadashi Miyasaka*

School of Pharmaceutical Sciences, Showa University, Hatanodai 1-5-8, Shinagawa-ku, Tokyo 142, Japan

ABSTRACT

Stereocontrolled addition of carbon radicals to 4',5'-unsaturated uracil-nucleosides can be accomplished by using organoselenium reagents as a radical source. This provides a new method for C-C bond formation at the 5'-position.

INTRODUCTION

We have recently reported on the reaction of a new phenylselenide anion, generated from $(PhSe)_2$ and $LiAlH_4$, with anhydro uracil-nucleosides, which leads a highly efficient method for introduction of a phenylselenenyl group to the sugar portion.[1,2] Selenoxide fragmentation of the resulting products readily furnished unsaturated-sugar nucleosides under mild conditions. Although these nucleosides seem to be attractive starting materials for C-C bond formation in the sugar portion, their use has so far been limited to the addition reaction of hetero atoms or hetero substituents.[3]

In connection with our previous studies,[4] we would like to report here that stereoselective addition of carbon radicals to 4',5'-unsaturated uracil-nucleosides can be accomplished by the use of organoselenium reagents as a radical source.

RESULTS

When the 2',3'-bis-O-tert-butyldimethylsilyl derivative (1) was treated with RSePh (R: CO_2Et, CH_2CO_2Et, CH_2COMe, CH_2CN) in the presence of Bu_3SnH and AIBN in refluxing benzene, exclusive formation of $C^{5'}$-substituted 1-(5-deoxy-α-L-lyxofuranosyl)uracils (2) was observed. In contrast to this, the 2',3'-di-O-acetyl

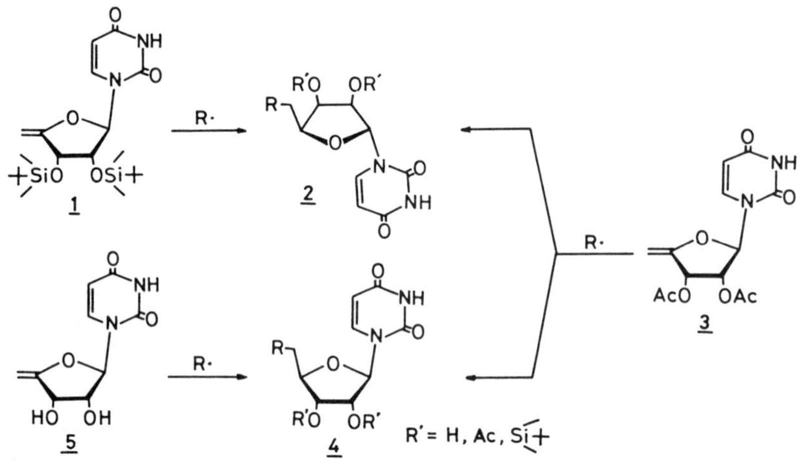

counterpart (3) gave, under similar reaction conditions to the above, a mixture of nearly equal amounts of two stereoisomeric products (2 and 4). Since the reaction is presumed to occur by initial attack of R· to the 5'-position to leave the $C^{4'}$-radical, we reasoned that these stereochemical outcomes might be the main reflection of steric hindrance imposed by the *tert*-butyldimethylsilyl or acetyl group during hydrogen abstraction of the $C^{4'}$-radical from Bu_3SnH.

We, therefore, next examined the use of 5. After several unsuccessful experiments, DMF was found to be a suitable solvent. Thus, when a DMF solution of 5 and RSePh was treated at 70 °C with a mixture of Bu_3SnH and AIBN in benzene, a highly stereoselective formation of 4 could be accomplished.

Application of this method to the 4',5'-unsaturated 2'-deoxynucleosides will also be discussed.

REFERENCES

1) Haraguchi, K., Tanaka, H., Hayakawa, H., Miyasaka, T. (1988) Chem. Lett., 931.
2) Haraguchi, K., Tanaka, H., Miyasaka, T. (1989) Synthesis, in press.
3) Verheyden, J. P. H., Moffatt, J. G. (1975) J. Am. Chem. Soc., 97, 4386.
4) Haraguchi, K., Tanaka, H., Hayakawa, H., Miyasaka, T. (1988) Nucleic Acids Symposium Series, 20, 33.

Synthesis of 2′-deoxy-6,2′-methano-pyrimidine nucleosides and optical properties of pyrimidine C-cyclonucleosides

Yuichi Yoshimura, Akira Matsuda, and Tohru Ueda

Faculty of Pharmaceutical Sciences, Hokkaido University, Sapporo 060, Japan

ABSTRACT

2'-Deoxy-6,2'-methano derivatives of uridine, cytidine, and 4-thiouridine were synthesized by the Peterson olefination of a ketosugar with a pyrimidine followed by an intramolecular glycosylation reaction. CD-spectra of these cyclonucleosides together with other pyrimidine C-cyclonucleosides were given.

To investigate stereochemical interactions of nucleosides or nucleotides with enzymes, the substrate fixed in a specific conformation seems to be very useful. We have developed various methods for the synthesis of carbon-bridged cyclonucleosides fixed in a certain glycosyl torsion angle. Recently we developed a new methodology of C-cyclonucleoside synthesis which employed condensation of an oxo-sugar with a lithiopyrimidine followed by the intramolecular glycosylation.[1] This method is now extended for preparation of 2'-deoxy-6,2'-methanouridine **6**.

2,4-Dimethoxy-6-trimethylsilylpyrimidine **1** was lithiated and then condensed with the pent-2-ulofuranoside **2** to give the (**Z**)-benzylidene derivative **3** in 76% yield under the Peterson-olefination conditions. Compound **3** was converted to the 1-*O*-acetate **4** by way of catalytic hydrogenation of **3** and manipulation of the protecting groups. Compound **4** was treated with $SnCl_4$ to afford the cyclonucleoside **5** , which was derivatized to 2'-deoxy-6,2'-methano-uridine **6**, -cytidine **7** and -4-thiouridine **8**, respectively.

The CD spectra of **6** and **8** were compared with those of other three cyclonucleosides[1-3] (Fig. 1 and 2). As the glycosyl torsion angles (C6-N1-C1'-O1') increase, the sign and magnitude of the Cotton bands of uridine derivatives (Fig 1) turn from positive to negative, the 6,6'-cyclo derivatives

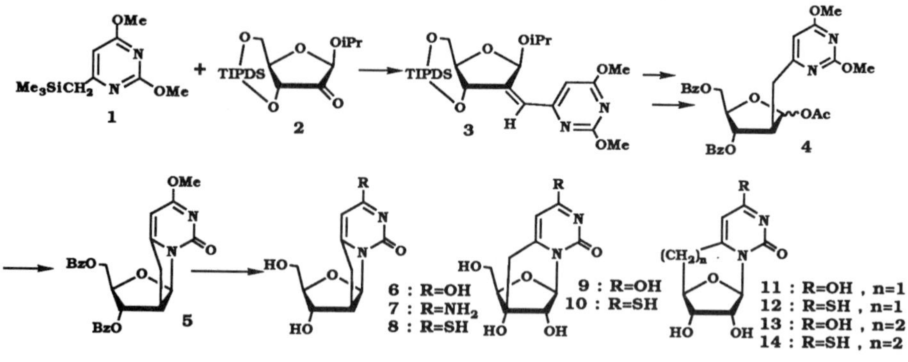

Fig. 1 CD spectra of cyclo-uridines

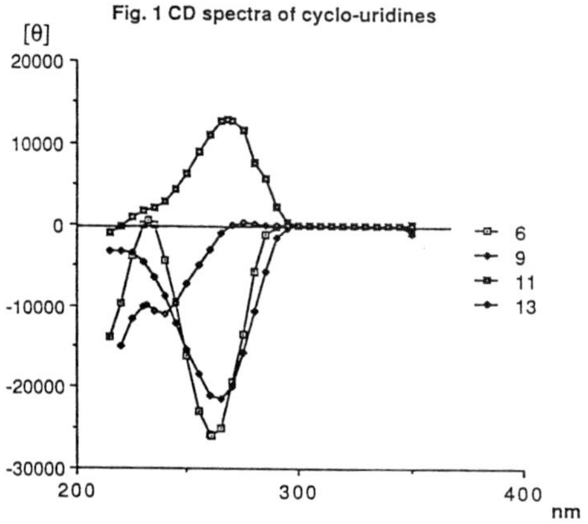

Fig. 2 CD spectra of cyclo-4-thiouridines

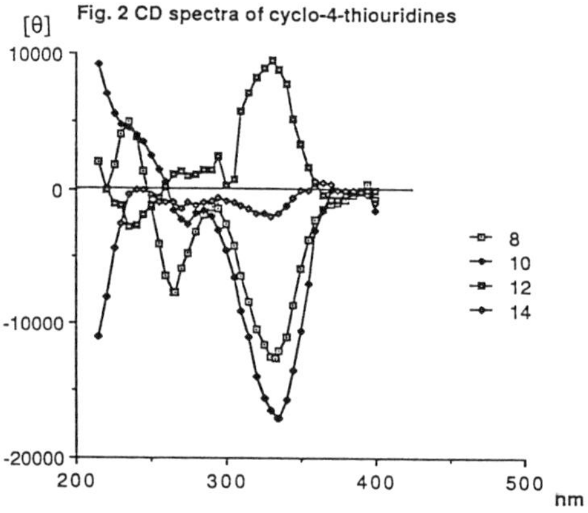

having the transitional torsion angle. By contrast, the 4-thiouridine counterpart **8** showed somewhat irregular band, which may be due to the different nature of the 4-thiouracil chromophore.

The interaction of these cyclonucleosides (phosphates) with pancreatic ribonuclease A will also be presented.

REFERENCES

1 Yoshimura, Y., Sano, T., Matsuda, A. and Ueda, T. (1988) *Chem. Pharm. Bull.*, **36**, 162.
2 Yoshimura, Y., Matsuda, A. and Ueda, T. (1989) *Chem. Pharm. Bull.*, **37**, 660.
3 Ueda, T., Usui, H., Shuto, S. and Inoue, H. (1984) *Chem. Pharm. Bull.*, **32**, 3410.

Synthesis and properties of the box 9R and 9R′ sequences of *Tetrahymena* rRNA

Shunichi Yamakage, Osamu Sakatsume, and Hiroshi Takaku*

Laboratory of Bioorganic Chemistry, Department of Industrial Chemistry, Chiba Institute of Technology, Tsudanuma, Narashino, Chiba 275, Japan

ABSTRACT

The chloroethoxyethyl (CEE) group is completely stable under the acidic conditions required to remove the 5'- protecting groups in the oligoribonucleotide synthesis, but can be cleaved under the similar condition to that of the tetrahydropyranyl (THP) group in region of pH 2-3. The oligoribonucleotides were synthesized by the phosphoramidite method on solid supports.

INTRODUCTION

Recent studies have indicated[1] that the tetrahydropyranyl (THP) and methoxytetrahydropyranyl(MTHP) groups were unstable under acidic conditions required for the removal of 5'-protecting groups in the solid phase synthesis of RNA sequences. In order to overcome this problem, several workers have already developed the new protecting groups for the 5'- and 2'-hydroxyl functions in the oligoribonucleotide synthesis on solid supports. In this paper, we wish to report that the CEE group is stable under acidic conditions required for the complete removal of a DMTr group; however, under mild hydrolytic condition (pH 2.0), the CEE group is removed.

RESULTS AND DISCUSSION

First, we examined the synthesis of uridine derivatives bearing three different 2'-acetal groups. Further, we carried out investigation without separation of the diastereoisomers. The reaction between 3',5'-O-(tetraisopropyldisiloxane-1,3-dilyl) uridine and alkyl vinyl ethers the conversion of the products obtained into **(2a-e)** are indicated in outline in scheme 1. The stability of 2'-acetal groups of uridine derivatives **(2a-e)** thus obtained were tested under acidic conditions such as 0.01N HCl and 1.5%

$Cl_2CHCOOH$. This result shows that **2e** was more stable under acidic conditions than other uridine derivatives **(2a-d)**. Additionally,

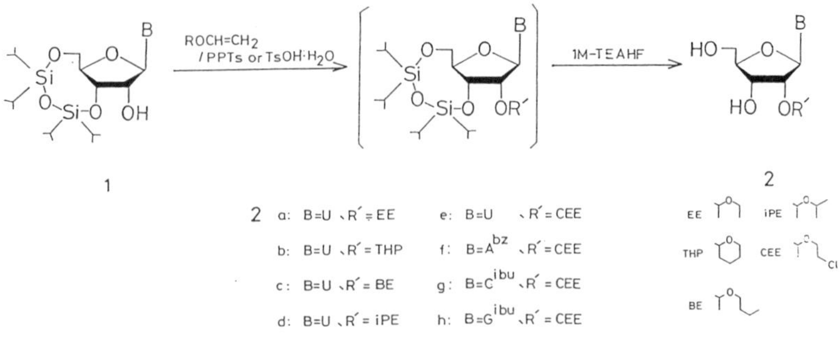

Scheme 1

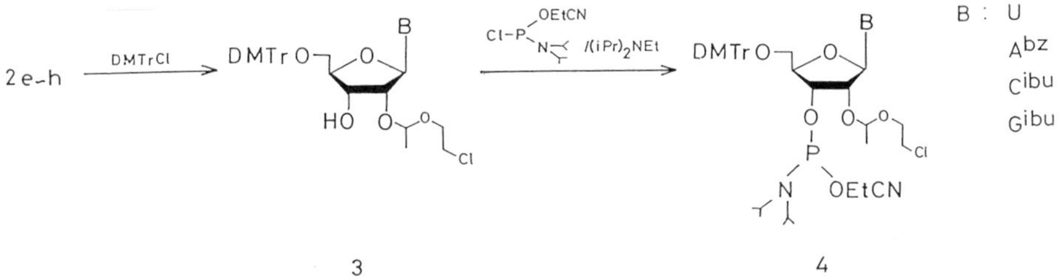

Scheme 2

2e has hydrolysis properties similar to that the THP or MTHP protecting groups relatively mild condition, for example, by treatment with 0.01N HCl (pH 2.0) at room temperature. In order to study of both internucleotide cleavage and phosphoryl migration under deprotecting conditions, the dimer, UpU was treated first with ammonia and then with 0.01 N HCl (pH 2.0). The internucleotide linkages was checked by the reversed phase HPLC analysis and enzymatic hydrolysis by ribonuclease T_2. No both internucleotide cleavage and phosphoryl migration were observed. Further, we have applied these nucleoside derivatives **(2e-h)** to the synthesis of Tetrahymena rRNA, box 9R (UGUCGGUC) and box 9R' (GACCGUCA). The coupling and deprotcting reactions were proceeded smoothly. The physical properties of the octaoligoribonucleotides, (box 9R and 9R') are now in progress.

REFERENCES

1. a) Reese, C. B. and Skone, P.A .(1985) Nucleic Acids Res., **13**, 5215-5231; b) Christodoulou, C., Agrawal, S. and Gait, M. J. (1986) Tetrahedron Lett., **27**, 1521-1522.

Self-assembly of synthetic deoxyoligonucleotides containing d-G clusters

Toshio Shida, Kohji Yokoyama, and Junichi Sekiguchi

Faculty of Textile Science and Technology, Shinshu University, Ueda, Nagano 386, Japan

ABSTRACT

Deoxyoligonucleotides containing d-G cluster were synthesized to elucidate exact structures and properties of the parallel-four stranded complexes of oligo(dG). Among these oligomers, d-TTGGGGTT and d-TTGGGGGGTT formed stable complexes which were able to interact with ethidium bromide known as an intercalator.

INTRODUCTION

Tetrameric molecular arrangements found in gels of guanosine derivatives are reproducible in the self-assembled structures of poly(G) and its 2-deamino analog poly(I)[1]. With respect to poly(dG), it is merely revealed that oligo(dG) protruding from 3' termini of plasmid DNA molecules can form several polymeric structures[2]. Such conformations derived from guanine-rich sequences seem remarkable in connection with any biological functions of gene promoters[3] and telomeres[4].

Here we report studies on the self-assembly of short deoxyoligomers containing d-G cluster.

RESULTS

Deoxyoligonucleotides which have two TT dimers on the both ends of d-G cluster to prevent stiky gel formation were synthesized. These sequences were similar to a part of telomeres.

Polyacrylamide (25%) gel electrophoresis of the oligomers are shown in Fig. 1. d-TT$(GG)_2$TT and d-TT$(GG)_3$TT samples stored in a refrigerator formed complexes, because they moved shorter distances than those anticipated from the mobilities of pigments and tetraoligonucleotide (lanes 1, 3, and 5). Sizes of the complexes of d-TT$(GG)_2$TT and d-TT$(GG)_3$TT were estimated to be approximately 30 Nts and 40 Nts, respectively, as a single strand. The complex of d-TT$(GG)_2$TT was changed to the single stranded form by heat denaturation, while d-TT$(GG)_3$TT complex was rather resistant to the heat denaturation. These data suggested that these complexes consisted of the parallel-four stranded structure.

The self-assembled d-TT$(GG)_2$TT showed a large positive CD band in the 250-290 nm region. This CD spectral pattern was very similar to that of the poly(G)[5]. Thus we note that the complex of d-TT$(GG)_2$TT is useful as a model compound of poly(dG).

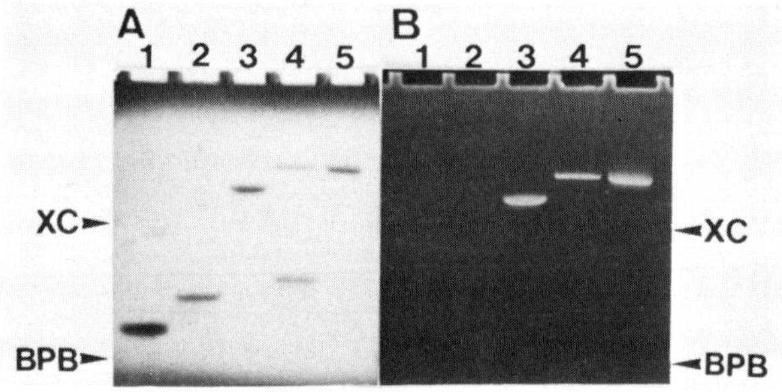

Figure 1. **Polyacrylamide gel** electrophoretic separation of the synthetic deoxyoligonucleotides. (A) DNA bands were detected by UV-shadowing. (B) Gel was stained with ethidium bromide. Lane 1, d-$A(pA)_3$; lanes 2 and 3, d-TT$(GG)_2$TT ; lanes 4 and 5, d-TT$(GG)_3$TT. The samples had been previously stored at a concentration of 20 A_{260} at 0°C in an aqueous solution. Prior to loading, the samples (lanes 2 and 4) were heated at 95°C for 5 min and then quenched in chilled water.

REFERENCES

1. Zimmerman,S.B., Cohen,G.H., and Davies,D.R. (1975) J.Mol.Biol., 92,181.
2. Dugaiczyk,A., Robberson,D.L., and Ullrich,A. (1980) Biochemistry, 19, 5869.
3. Sen,D. and Gilbert,W. (1988) Nature, 334, 364.
4. Henderson,E., Hardin,C.C., Walk,S.K., Tinoco,Jr,J., and Blackburn,E.H. (1987) Cell, 554, 899.
5. Howard,F.B., Frazier,J., and Todd Miles, H. (1977) Biopolymers, 16, 791.

Electronic structure of DNA by DV-Xα cluster calculations: II. d(GG)·d(CC), d(CG)$_2$, d(GC)$_2$ A and B conformations. (Part 2) sugars and bases

Takako Shinoda, Nobuyuki Shima[1], and Masaru Tsukada[1]

National Chemical Laboratory for Industry, Tsukuba Research Center, Tsukuba, Ibaraki 305 and [1]Department of Physics, Faculty of Science, University of Tokyo, 7-3-1 Hongo, Bunkyo-ku, Tokyo 113, Japan

ABSTRACT

The electronic structure of d(GG)·d(CC), d(CG)$_2$, d(GC)$_2$ which are stacked base pairs in the DNA double helix, are elucidated for both A and B conformations in detail by DV-Xα cluster calculations. These three DNA double helix fragments are constracted from the same bases, G and C, but the electronic structures of the fragments for both A and B conformations are different from each other characteristically. There are some delicate differences in the admixture of the orbital components and the overlap populations of intra- and inter- strand stacked bases among the stacking isomers. On the other hand, the electronic states of sugars differ in the 5'→3' direction, but are not almost dependent on stacked base pairs.

INTRODUCTION

The drastic differences in the electronic structure of phosphates among these stacking isomers were reported previously [1]. In particular, the differences in the level structures of the O2, O3 of each phosphate make capable of perfect recognition of the kind of the stacked base pair together with the 5'→3' direction. In this report we focus our attention on bases and sugars in the electronic structures of GG/CC, CG/GC, GC/CG clusters in both A and B forms.

METHOD

The application of the DV-Xα cluster method and the geometries of cluster models have been fully described in the previous report [1].

RESULTS AND DISCUSSION

In all clusters the LUMO (291a) and the level 292a just above it are made mainly of the π-electron orbital of cytosine.

On the other hand, the HOMO (290a) is made mainly of the 2p of the O2 and O3 of a phosphate. The HOMO in the partial density of states (DOS) of bases locates at the energy position just below the eight levels which are mainly composed of the 2p states of the O2 and O3 atoms of each phosphate. The energy and the orbital componrnts of the 291a, 292a, and the HOMO of the base-DOS and the level just below it, which are made mainly of the π-electron orbital of guanine, are listed in Table for B forms. From Table, it is noted that no admixture of the π-orbitals of the stacked bases are found in the 291a, 292a for all clusters and in the 282a, 281a for the GC/CG cluster, whereas the π-orbitals of guanine of GG/CC and CG/GC clusters admix between the stacked G-bases. The differences among the stacking isomers appear also in the overlap populations of intra- and inter- strand stacked bases. We might successfully explain on the basis of the results the experimental facts that DNA strand scission mediated by BLM occurs at 3'-C of G$\rightarrow$C strand and BLM attacks the N2 of guanine.

The calculated DOS's of sugars locate in the deep energy regions, $|E|>2\sim4$ eV, and energy shifts below 0.6 eV are found among the stacking isomers, whereas the shape of DOS differs in the 5'$\rightarrow$3' direction regardless of the kind of the binding base.

TABLE

	291a			292a		
	GG/CC(B)	CG/GC(B)	GC/CG(B)	GG/CC(B)	CC/GC(B)	GC/CG(B)
Energy(eV)	0.05	0.41	0.51	0.31	0.53	0.63
Base-1	-	0.975	0.003	-	-	0.001
Base-2	-	-	0.002	-	0.002	0.970
Base-3	0.968	-	0.001	0.010	0.978	0.002
Base-4	0.011	0.003	0.969	0.963	-	0.002
Others	0.017	0.019	0.018	0.018	0.017	0.019

	284a	282a		282a	281a	
	GG/CC(B)	CG/GC(B)	GC/CG(B)	GG/CC(B)	CG/GC(B)	GC/CG(B)
Energy(eV)	-1.71	-1.92	-2.06	-1.95	-2.11	-2.18
Base-1	0.753	-	0.964	0.244	0.002	0.012
Base-2	0.227	0.210	0.007	0.742	0.757	0.001
Base-3	-	0.001	0.017	-	-	0.970
Base-4	-	0.754	-	-	0.217	0.008
Others	0.012	0.026	0.011	0.011	0.020	0.004

REFERENCE [1] T. Shinoda, N. Shima, M. Tsukada, 58th Annual Meeting of the Chem. Soc. Jpn, 1989, 3IIG03.

Symposium Series No. 21 1989 Nucleic Acids Research

Spontaneous formation of helical strands from nucleic acid–lipid conjugates

Hiroshi Yanagawa*, Yoko Ogawa, Hiroyuki Furuta, and Katsushige Tsuno[1]

Mitsubishi Kasei Institute of Life Sciences, 11 Minamiooya, Machida, Tokyo 194 and [1]JEOL Ltd., 1418 Nakagami, Akishima, Tokyo 196, Japan

ABSTRACT

We have developed a new class of helical strands that self-assembled spontaneously in aqueous solution. Phospholipid-deoxy-nucleoside conjugates containing two myristoyl groups and a deoxy-nucleosidyl group spontaneously assembled to form different types of helical strands.

Molecular helicity is the most fundamental property displayed by biological polymers such as nucleic acids, proteins, and starch. DNA, which is a genetic store, consists of two right-handed helical polynucleotides formed by linking 3'- and 5'-position of adjacent sugar residues by a phosphodiester bond. Phospholipids, which are quantitatively dominant constituents in biological membranes, self-assemble to form micelles, monolayers, and bilayer vesicles in aqueous media. A mononucleotide unit noncovalently linked through the hydrophobic groups of lipids should be able to self-assemble in aqueous media and form helical strands like DNA. Recently we have discovered that phospholipid-nucleoside conjugates spontaneously assemble to form linear and circular helical strands.[1-3] We concluded that stacking and hydrogen bonding between bases, and hydrophobic interactions between the long alkyl chain moieties of phospholipid-nucleoside conjugates are necessary for the formation of the helical strands. We expected to construct different helical structures by controlling stacking and hydrogen bonding between bases of phospholipid-nucleoside conjugates. Consequently, we have synthesized dimyristoyl-5'-phosphatidyldeoxynucleosides with different nucleic acid bases.

Dimyristoyl-5'-phosphatidyldeoxynucleosides were enzymatically

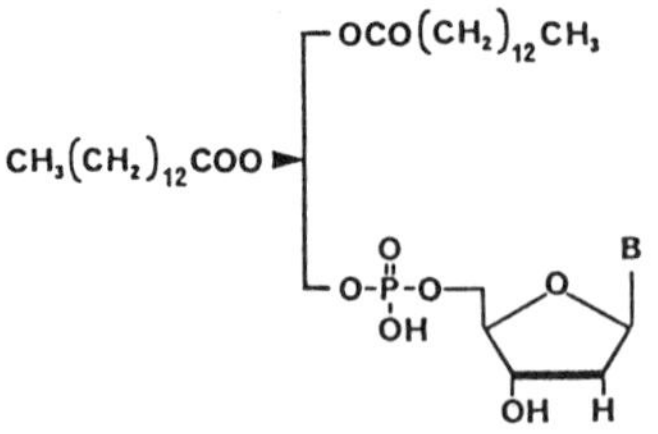

Dimyristoyl-5'-phosphatidyldeoxynucleoside

(DMPdN)

B =

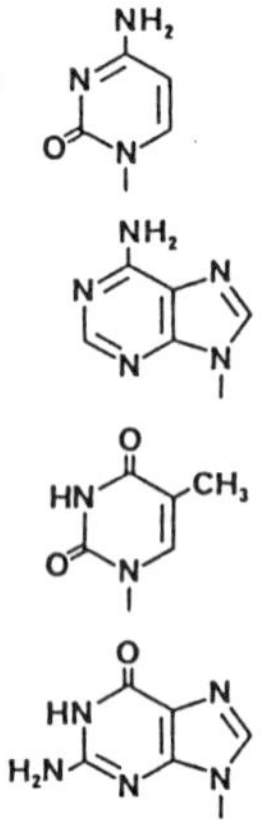

Cytosine (DMPdC) 1

Adenine (DMPdA) 2

Thymine (DMPT) 3

Guanine (DMPdG) 4

synthesized from 1,2-dimyristoyl-sn-glycero-3-phosphocholine and corresponding deoxynucleoside.[4] The helical strands of dimyristoyl-5'-phosphatidyldeoxynucleosides were prepared in aqueous solution, and examined under an electron microscopy and by automatic image processing as described previously.[1]

After sonication at 50°C for 45 min, an aqueous solution of **1** produced vesicles mainly with diameters of 200-1600Å. The vesicles were slowly transformed into two types of helical strands, i.e., a thin and thick helical strands after aging at 25°C for 1 day. Image processing showed that the thin helical strand consisted of a duplex with a diameter of ≈110Å and a helical pitch of ≈240Å. The duplex further formed a superhelical structure with a helical pitch of ≈1100Å. The helix was right-handed. Furthermore, **2**, **3**, and **4** formed characteristic helical strands in aqueous solution, respectively.

We thank to Dr. S. Shuto of Toyo Jozo Co., Ltd. for providing Streptomyces phospholipase D.

REFERENCES

1. Yanagawa, H., Ogawa, Y., Furuta, H. & Tsuno, K. (1988) Chem. Lett., 269-272.
2. Yanagawa, H., Ogawa, Y., Furuta, H. & Tsuno, K. (1989) Chem. Lett., 403-406.
3. Yanagawa, H., Ogawa, Y., Furuta, H. & Tsuno, K. (1989) J. Am. Chem., Soc., in press.
4. Shuto, S., Ueda, S., Imamura, S., Fukukawa, K., Matsuda, A. & Ueda, T. (1987) Tetrahedron Lett. 28, 199-202.

ESR behaviour of spin-labelled oligoDNAs

Akira Murakami, Shunji Nagahara, Tamio Takeuchi, and Keisuke Makino

Department of Polymer Science and Engineering, Faculty of Textile Science, Kyoto Institute of Technology, Matsugasaki, Sakyo-ku, Kyoto 606, Japan

ABSTRACT

Oligodeoxyribonucleotides (oligoDNAs) were spin-labelled at the 5'-end internucleotide linkage, directly or through the spacers of different lengths, with TEMPO and the labelled oligoDNAs were explored for the ESR spectroscopical behaviour in solution. Oxidation of H-phosphonate intermediates in CCl_4 solutions of TEMPO and diamines was taken for the introduction of TEMPO and spacers, respectively. ESR lines of TEMPO attached to oligoDNAs, directly or through spacers, were already broadened compared to those of free TEMPO, and the broadening decreased with increasing chain length of the specer and increased with increase in the chain length of the oligoDNAs. The ESR lines of the oligoDNA labelled directly were further broadened in the presence of a DNA complementary to the labelled oligoDNA. In addition, the broadened lines were further broadened in the presence of a larger DNA. This observation implies that spin labelling at the internucleotide linkage is useful in the DNA probe method to avoid B/F separation.

INTRODUCTION

DNA probe diagnoses for detecting genetic defects and point mutations are important in medicine. There are two major problems in DNA-probe diagnosis. First, target DNA must be fixed on the appropriate matrices such as nitrocellulose membrane. Secondly, DNA-probe must be carefully washed after hybridization to minimize the non-specific binding on the membrane. To overcome these problems, we have studied usage of spin-labels which can be detected in solution by ESR without B/F separation. In the present study , oligodeoxyribonucleotides (oligoDNAs) site-specifically spin-labelled with 2,2,6,6-tetramethylpiperidine-N-oxyl (TEMPO) were prepared by means of the hydrogen phosphonate method and explored for the ESR spectroscopical behaviour in solution.

RESULTS AND DISCUSSION

Two types of spin-labelled oligoDNAs were prepared. Directly labelled oligoDNAs were prepared by previously described procedure[1]. For the preparation of oligoDNAs labelled through spacers, diamines (ethylenediamine and hexamethylenediamine) were first introduced to the internucleotide linkage by the same procedures, followed by introduction of 4-isothiocyanate-TEMPO in 0.1 M carbonate-formamide buffer (pH9.0).

ESR lines of TEMPO attached to oligoDNAs, directly or through spacers, were already broadened compared to those of 4-amino-TEMPO, indicating that the mobility of TEMPO was restricted by the introduction to oligoDNAs. When the chain length of oligoDNAs are same, the broadening of the ESR lines decreased as the length of the spacers increased. When TEMPO was introduced to oligoDNAs directly, the broadening of the ESR lines increased with increase in the chain length of oligoDNAs. These results indicate that the way of introducing TEMPO to oligoDNAs is largely responsible for the resultant ESR spectra and that the spin-label attached to oligoDNAs is an effective indicator for the behaviour of the spin-labelled oligoDNAs in solution.

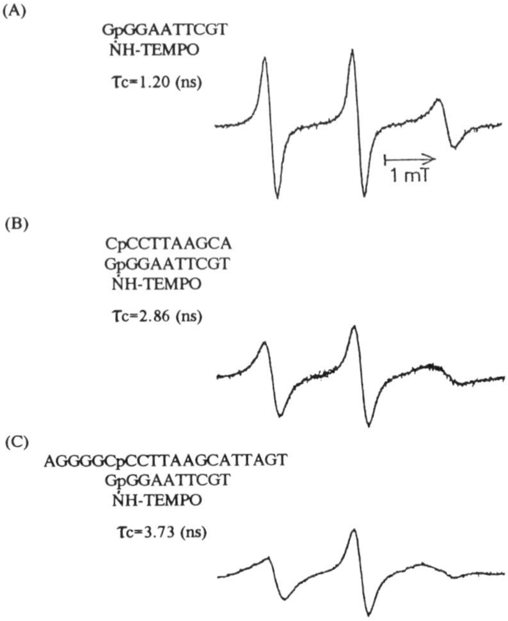

Fig.1 Change of the ESR spectrum of the spin-labelled oligoDNA in the presence of complementary oligoDNAs.

In the present study, oligoDNAs directly labelled with TEMPO were used to see if duplex formation can be observed by use of this DNA probe.

When the spin-labelled oligoDNA probe, shown in Fig.1(A), was mixed with its complementary oligoDNA (10mer), further broadening of the ESR lines was observed, as represented in Fig.1(B). The correlation times (τ_c from Kivelson's equation) are also accompanied in the figures. Above results show the duplex formation between the oligoDNA and the spin-labelled oligoDNA. The line broadening due to the duplex formation was further examined by measuring this probe in the presence of a larger oligoDNA (20mer). As depicted in Fig.1(C), the ESR lines were further broadened, supporting above cosiderations.

Results obtained here indicate that by use of spin-labelled oligoDNAs duplex formation can be clearly detected by ESR and that the detection of target DNA in solution can be done without B/F separation.

REFERENCES

1. a) K.Makino, A. Murakami, S. Nagahara, Y. Nakatsuji and T. Takeuchi (1988) Nucleic Acids Symposium Series, 20, 89-90. b) K. Makino, A. Murakami, S. Nagahara, Y. Nakatsuji and T. Takeuchi (1989) Free Radical Research Communications, in press.

Uranyl-ion catalyzed oligonucleotide synthesis from adenosine-5′-phosphorazolide. The effect of azoles as nucleotide activating groups

Hiroaki Sawai, Takashi Shibusawa, and Kensei Kuroda

Department of Chemistry, Faculty of Engineering, Gunma University, Kiryu, Gunma 376, Japan

ABSTRACT

Polymerization of azole-activated nucleotides was conducted in an aqueous medium with uranyl ion as a catalyst. Imidazole, 2-methylimidazole, 4-methylimidazole, triazole, nitrotriazole, tetrazole and 2-methylbenzimidazole were used as an activation group for adenylic acid. Oligoadenylates up to the hexadecamer were obtained in high total yield under the optimum condition. The azole compounds had large effect on the rate of polymerization, the yield and the chain length of the resulting oligoadenylates

Recently we have reported that uranyl ion catalyzes the polymerization of nucleoside-5'-phosphorimidazolide very efficiently to yield oligoadenylates[1]. The resulting oligomers contain mainly 2'-5' internucleotide linkage, though the 2'-5' regio-selectivity varied depending on the reaction condition. Inoue and Orgel reported the remarkable enhancement of the 3'-5' regio-selectivity and the improvement of the chain length in the template-directed oligoguanylates synthesis on poly C, by replacing guanosine-5'-phosphorimidazolide with guanosine-5'-phosphor-2-methylimidazolide as a starting monomer[2]. The effect of the substituent group of imidazole on the polymerization is very interesting. In this study, we carried out uranyl-ion catalyzed polymerization of various adenosine-5'-phosphorazolide to examine the effect of nucleotide activating azoles on the oligoadenylate formation, and to control the chain length and the regio-selectivity of the resulting oligoadenylates.

Adenosine-5'-phosphorazolide(AzpA) was prepared from

adenylic acid and azoles; imidazole, 2-methylimidazole, 4-methylimidazole, triazole, 3-nitrotriazole, tetrazole and 2-methylbenzimidazole. Polymerization of AzpA was conducted in a reaction mixture containing AzpA(0.05 M), uranyl nitrate (5mM~10μM) and N-ethylmorpholine buffer(0.2M, pH7.3) at room temperature. The reaction mixture was treated with EDTA-solution and analyzed by HPLC. Uranyl ion catalyzes the polymerization of adenosine-5'-phosphorimidazolides very efficiently forming oligoadenylates up to the hexadecamer. The 2'-5' internucleotide linkage was preferentially formed in the reaction. Replacement of imidazole with 2-methylimidazole or 4-methylimidazole as an activating group resulted in a small decrease of 2'-5' regio-selectivity and in a small increase of the chain length of the oligoadenylates under the same condition. The steric hindrance of 2-methylbenzimidazole suppressed both polymerization and hydrolysis of adenosine-5'-phosphor-2-methylbenzimidazolide largely. However, high reaction temperature accelerate its reaction. The polymerization was almost complete in 4 h at 50 °C giving 2'-5' linked oligoadenylates from dimer to heptamer in more than 55 % total yield.

No long-chained oligoadenylates could be obtained from adenosine-5'-phosphor triazolide(TrpA). 2'-5' linked dimer, trimer and tetramer were formed from TrpA in 18, 32 and 10 % yield when 1 mM of uranyl ion was used as a catalyst. No efficient polymerization took place from nitrotriazolide and tetrazolide of adenylic acid. Hydrolysis of phosphorazolide bond and pyrophosphate formation were predominant in the cases. In conclusion, azole compounds as an activating group of adenylic acid have large effect on the rate of the polymerization and on the chain length of the resulting oligoadenylates. However, they have very little effect on the regio-selectivity of the internucleotide linkage.

REFERENCES

1) H. Sawai, K. Kuroda and T. Hojo, Bull. Chem. Soc. Japan, 62, No 7 (1989)
2) T. Inoue and L. E. Orgel, J. Amer. Chem. Soc., 103, 7666 (1989)

8-Methyladenosine-substituted analogs of 2–5A: synthesis and properties

Yukio Kitade, Yoshitaka Nakata, Kosaku Hirota, Yoshifumi Maki, Aysun Pabuccoglu[1], and Paul F.Torrence[1]

Medicinal Chemistry, Gifu Pharmaceutical University, 5-6-1 Mitahora-higashi, Gifu 502, Japan and [1]Biomedical Chemistry, Laboratory of Analytical Chemistry, NIDDK, National Institutes of Health, Bethesda, MD 20892, USA

ABSTRACT

Four 8-methyladenosine-substituted analogs of 2-5A were obtained by a synthesis which involved a modification of the lead ion-catalyzed ligation reaction. Substituent of an 8-methyladenosine residue at the third position (2'-terminal) of the oligonucleotides increased the stability to snake venom phosphodiesterase digestion.

INTRODUCTION

The proposed role of the 2-5A system in the mecanism of interferon's antiviral action and in cell regulation or differentiaion suggests its use as a new approach to the chemotherapy of viral or neoplastic diseases.[1] However, such applications of 2-5A are limited by at least two considerations; first, 2-5A has a short biological half-life due to rapid degradation by 2',5'-phosphodiesterase: secondly, owing to its highly polar nature, 2-5A does not penetrate the eukaryotic cell.[2] Recently we have found that the diphosphate, pp5'A2'p5'A2'p5'(br^{8}A), which although binding to endonuclease slightly less effectively than 2-5A, was 10 times more effective as an inhibitor of translation. In contrast to the "tailed" analogs of 2-5A its increased stability toward degradation does not seem to be the explanation for its increased activity.[3] Endonuclease activation might require that the third adenosine residue adopts a *syn* glycoside conformation. This assumption prompts research on the synthesis of the other 8-substituted adenosine (i.e. 8-methyladenosine) analogs of 2-5A and their properties.

RESULTS AND DISCUSSION

8-Methyladenosine was obtained with ease by reaction of 8-bromoadenosine with $Al(CH_3)_3$ in the presence of Pd-catalyst.[4] Treatment of 8-methyladenosine with $POCl_3$ in $(CH_3O)_3PO$ gave the corresponding 5'-monophosphate [p(me^{8}A)] quantitatively. The corresponding 8-methyladenosine-substituted trimer 5'-monophosphates, i.e. p5'A2'p5'A2'p5'(me^{8}A), p5'A2'p5'(me^{8}A)2'p5'(me^{8}A), p5'(me^{8}A)2'p5'(me^{8}A)2'p5'(me^{8}A), and p5'(me^{8}A)2'p5'A2'p5'A, were synthesized by employment of a modified lead-catalyzed ligation reaction.[5] These trimer 5'-monophosphates were further phosphorylated to the corresponding 5'-triphosphates by the reaction with pyrophosphate using N,N'-carbonyldiimidazole as a condensing agent. The structures of these 2-5A analogs were supported by HPLC analyses (BAP, SVP, and BAP-SVP digestions and alkaline degradation) and NMR spectral data.

When snake venom phosphodiestrase was employed for the preliminary digestion of the trimer 5'-monophosphates instead of cell extract, introduction of 8-methyladenosine residue at the third position (2'-terminal) of the oligonucleotides such as p5'A2'p5'A2'p5'(me^{8}A) and p5'(me^{8}A)2'p5'(me^{8}A)2'p5'(me^{8}A) increased their stability to SVP digestion. The diphosphate p5'(me^{8}A)2'p5'A2'p5'A, however, was more quickly degraded than 2-5A monophosphate itself.

Evaluation for their ability to activate the 2-5A dependent endonuclease by means of a cell-free protein synthesis is now in progress.

ACKNOWLEDGEMENTS

This work was supported, in part, by the Grant-in-Aid for Scientific Research from the Ministry of Education of Japan, which is gratefully acknowledged.

REFERENCES

1. Torrence, P. F. (1982) Mol. Aspects Mod. 5, 129.
2. Torrence, P. F. (1985) In Torrence, P. F. (ed.), Biological Response Modifiers, Academic Press, Inc., New York, pp 77-105.
3. Lesiak, K. and Torrence, P. F. (1987) J. Biol. Chem. 262, 1961-1965.
4. Hirota, K., Kanbe, Y., KItade, Y., and Maki, Y. (1988) Nucleic Acids Res. Symposium Ser., 20, 31-32.
5. Imai, J. and Torrence, P. F. (1985) J. Org. Chem. 50, 1418-1426.

Structure and mechanism of the hammerhead self-cleaving domain

Olke C.Uhlenbeck, SueAnn C.Dahm, Duane E.Ruffner, and Martha J.Fedor

Department of Chemistry and Biochemistry, University of Colorado, Boulder, CO 80309-0215, USA

As part of their replication pathway, the genomes of several plant satellite RNAs are capable of spontaneous cleavage at a unique site to give 2',3' cyclic phosphate and 5' hydroxyl termini (1, 2). The information needed for this reaction is contained in less than 40 nucleotides folded in a secondary structure (called the hammerhead) containing three helices and thirteen conserved nucleotides. We have studied this cleavage reaction by assembling the hammerhead with two synthetic oligoribonucleotides in two different ways (3, 4):

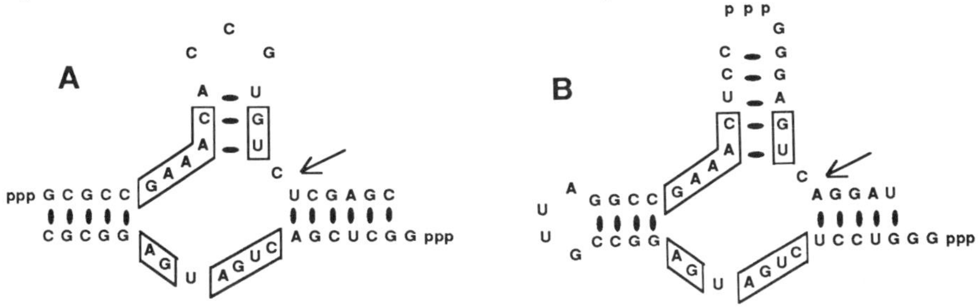

In both cases cleavage occurs at the expected site in the substrate oligoribonucleotide when complementary base pairs can form in the helical regions. Since many moles of substrate strand can be cleaved per mole of enzyme strand, true catalysis occurs. A kinetic study of several different B-type hammerheads indicates that while K_m varies considerably, k_{cat} is always about 2 min^{-1}. It appears likely that the large variability in K_m is a result of aggregation or alternate conformations of either oligomer reducing the rate of assembly of the hammerhead.

A thorough mutagenic analysis of hammerhead A indicates that 11 of the conserved nucleotides are essential for cleavage, presumably because they either participate in the mechanism of the reaction or are required for further folding to form a tertiary structure. Attempts to find potential tertiary interactions by looking for second site revertants were unsuccessful. Chemical modification and enzymatic digestion experiments provide additional evidence that the conserved nucleotides participate in further structure. Substitution interference experiments using α-thiophosphates indicate that several phosphates participate in the structure as well.

Hammerheads of the B-type were used to investigate the role of metal ions on the cleavage reaction. Although cleavage can occur in 1 M NaCl and 1 mM EDTA, the rate is greatly stimulated by divalent ions. A substrate containing a single thiophosphate at the cleavage site cleaves ~180-fold slower than phosphate in buffers containing $MgCl_2$, suggesting that cleavage rather than assembly is the rate limiting step of the reaction. However in buffers containing $MnCl_2$, the thiophosphate substrate cleaved at a normal rate. This clearly indicates that a divalent ion interacts with the phosphate at the cleavage site. Experiments using $MgCl_2$, $MnCl_2$ and the thiophosphate substrate indicate that divalent ions aid assembly of the hammerhead as well as catalysis.

REFERENCES

1. Prody, G.A., Bakos., J.T., Buzayan, J.M., Schneider, I.R. and Bruening, G. (1986) Science, 231, 1577-1580.
2. Hutchins, C.J., Rathjen, P.D., Forster, A.C. and Symons, R.H. (1986) Nucl. Acids Res., 14, 3627-3640.
3. Uhlenbeck, O.C. (1987) Nature, 328, 596-600.
4. Ruffner, D.E., Dahm, S.C. and Uhlenbeck, O.C. (1989) Gene, in press.

Synthesis of 2-deoxy-D-ribofuranosyl 5-amino- and 5-hydroxyaminoimidazole nucleosides by chemical methods and biotransformation

David Ewing, Robert W.Humble[1], Antonin Holy[2], Grahame Mackenzie[1], Gordon Shaw[3], and Ivan Votruba[2]

University of Hull, Hull HU6 7RX, [1]Humberside College of Higher Education, Hull HU6 7RT, UK, [2]Institute of Organic Chemistry and Biochemistry, Czechoslovak Academy of Sciences, 16610 Praha 6, Czechoslovakia, and [3]University of Bradford, Bradford BD7 1DP, UK

In our studies aimed at the synthesis of inhibitors of enzymes involved in the de novo biosynthesis of purine nucleotides, the nucleotite 5-amino-1-β-D-arabinofuranosylimidazole-4-carboxylate 5'-O-phosphate, which is analogous to CAIR, a central intermediate in the pathway,showed [1] significant inhibition and substrate activities. We therefore wished to synthesise the related nucleotide 5-amino-2'-deoxy-β-D-erythro-pentofuranosylimidazole-4-carboxylate 5'-O-phosphate for further enzymological studies. Additionally, we were also intersted to synthesise 5-hydroxy-2'-deoxy-β-D-erythro-pentofuranosylimidazole-4-carboxylate 5'-O-phosphate analogous to the immunosuppressive drug bredinin.

Previously reported routes to 2-deoxy-D-ribofuranosyl-aminoimidazole nucleosides have involved either degradation of purine nucleosides [2] or condensation [3] of 2-deoxyribofuranosyl halides with, such as, the nitro imidazole ester, methyl 5-nitroimidazole-4-carboxylate followed by hydrogenation.

We have earlier reported [4] the synthesis of ethyl 5-amino-1-2',3',5'-tri-O-benzyl-β-D-arabinofuranosylimidazole-4-carboxylate by direct condensation of ethyl 5-aminoimidazole-4-carboxylate with 2,3,5-tri-O-benzyl-α-D-arabinofuranosyl chloride in hot acetonitrile containing triethylamine. Using the same solvent conditions and ethyl 5-aminoimidazole-4-carboxylate we performed a comparable reaction using 1-chloro-2-deoxy-3,5-di-O-p-toluoyl-α-D-erythro-pentofuranose [5]. NMR analysis of the reaction mixture revealed two α and β-anomeric pairs of the N^1 and N^3 glycosylated imidazole in the yields 11 % α-N^1, 3.8 % β-N^1, 30 % α-N^3 and 9.2 % β-N^3 respectively. The isomers were only partly separated by silica gel column chromatography into two fractions containing the N^1 and N^3 isomers respectively; the anomeric mixtures in each case were not separated.

The stereospecific sodium salt glycosylation procedure for the synthesis of 2'-deoxyribonucleosides has been used [6] to synthesise a variety azole nucleosides. We therefore applied this approach by reacting the sodium salt of ethyl 5-aminoimidazole-4-carboxylate, produced in situ by NaH in CH_3CN, with 1-chloro-2-deoxy-3,5-di-O-p-toluoyl-α-D-erythro-pentofuranose. When we used NaH suspended in oil (60 % w/w) only modest yields were detected, by NMR analysis of column chromatography fractions, with poor selectivity towards β-anomers: 3.4 % α-N^1 , 1.5 % β-N^1 , 7.1 % α-N^3 , 2.3 % β-N^3. However, when anhydrous NaH was employed the β-isomers predominated but with the β-N^3 isomer being produced in the highest yield: undetected α-N^1, 10 % β-N^1, 1.3 % α-N^3 , 28 % β-N_1. The β-N^1 (m.p. 172-3°) and β-N^3 (m.p. 80-81°) crystallised readily following silica gel column chromatography.

Ethyl-5-hydroxyimidazole-4-carboxylate was synthesised in by reaction of diethylaminomalonate and formamidine acetate. Condensation of 1-chloro-2-deoxy-3,5-di-O-p-toluoyl-α-D-erythro-pentofuranose with the sodium salt of ethyl-5-hydroxyimidazole-4-carboxylate in CH_3CN yielded a mixture of both N- and O-glycosylated products.

We extended our investigation of the synthesis of 2-deoxy-β-D-ribofuranosyl-imidazole nucleosides by applying a biotransformation technique developed by Holy and Votruba.[7] The catalyst was composed of auxotroph thymine dependent Escherichia coli encapsulated in permeable alginate gel, and 2'-deoxyuridine, was the glycosyl donor Such conditions were found to glycosylate ethyl-5'-aminoimidazole-4-carboxylate to give an isomeric mixture of the two nucleosides ethyl-5'-amino-2'-deoxy-β-D-erythro-pentofuranoseimidazole-4-carboxylate (18 % β-N^1) and ethyl-4-amino-2'-β-D-erythro-pentofuranoseimidazole-5-carboxylate (35 % β-N^3) which were separated by HPLC [(6 % acetonitrile in 0.05 M triethylammonium hydrogen carbonate pH 7.5 (200 x 4 separon SG x 6 m)]. In contrast no reaction was observed with other azoles studied, namely, AICA, 4-cyano-5-hydroxyimidazole and 3-amino-4-ethoxycarbonylpyrazole. The ambiguity observed in the site of deoxyribosylation, by this biotransformation method, is contrary to results obtained [7] with pyrimidines and purines, so far studied, in which substitution takes place at sites N^1 and N^9 respectively.

References

1. G. Mackenzie, G. Shaw and S. E. Thomas, J.C.S. Chem. Comm., 1976, 454
2. J.A. Montgomery and H.J. Thomas, J. Med. Chem., 1972, 15, 182.
3. L. B. Townsend in "Synthetic Procedures in Nucleic Acid Chemistry", Vol. II, Eds. W.N. Zorbach and R. S. Tipson, Wiley-Interscience 1973.
   R.J. Rousseau, R. K. Robins and L.B. Townsend, J. Het. Chem., 1970, 7, 367.
4. G. Mackenzie and G. Shaw, J.C.S. Chem. Comm., 1978, 882.
5. C.C.Bhat in "Synthetic Procedures in Nucleic Acid Chemistry" Vol.I, Eds. W.W. Zorbach and R.S. Tipson, Wiley-Interscience 1968.
6. G.R. Revanker, K. Ramasamy and R.K. Robins, Nucleosides, Nucleotides, 1988, 7(3), 385
7. A. Holy and I. Votruba, Nucleic Acids Research, Symposium Series, 1987, 18, 69.

Attempted stereoselective synthesis of P-chiral analogues of oligodeoxyribonucleotides

Maria Jaworska, Zbigniew J.Leśnikowski, and Wojciech J.Stec

Polish Academy of Sciences, Centre of Molecular and Macromolecular Studies, Department of Bioorganic Chemistry, 90-363 Łódź, Sienkiewicza 112, Poland

Abstract

Reaction of diastereomeric 5'MMT-nucleoside 3'-O-(4-nitrophenylmethanephosphonate) or 3'-O-[O-(4-nitrophenyl)-S-(2-nitrobenzyl)phosphorothioate] with tBuMgCl-activated, 3'-protected nucleoside is stereospecific and leads, after appropriate work-up, to dinucleoside-3',5'-methanephosphonate or dinucleoside-3',5'-phosphorothioate, respectively. This procedure extended to 5'-activated nucleotides, allows to prepare short, stereoregular oligonucleotide analogues bearing at internucleotide positions methanephosphonate or phosphorothioate function of predetermined sense of chirality at P-centers.

Methanephosphonate and phosphorothioate analogues of oligodeoxyribonucleotides prepared by classical methods consist of the random mixture of 2^n diastereomers, where n is a number of internucleotide P-chiral linkages. It is conceivable that many of the stereomeric forms are unable to hybridize with the complementary oligonucleotide sequences by Watson-Crick base pairing interactions, whilst the remainder represent a range of hybridizing efficiencies depending upon their overall structure.

Recently we have reported on the stereoselective synthesis of tetrathymidylic acid analogues, bearing at internucleotide position methanephosphonate functions of all-R_P- and all-S_P-configurations[1]. Both compounds were characterized by means of ^1H and ^{31}P NMR and HPLC techniques. Condensation of 5'MMT($T_{PMe}T_{PMe}T_{PMe}T$) with ($T_{PMe}T_{PMe}T_{PMe}T$)3'-Ac by means of methanephosphonodichloridate led to octa(thymidine methanephosphonate) with all-except-one R_P-(1) and all-except-one-S_P-(2)-configurations at phosphorus. The presence of MMT group at 5'-end and -Ac group at 3'-end was proved by means of NMR and chemical transformations. Hybridization of octathymidylic acid (3) with pentadecadeoxyadenylic acid (dA_{15}) (4) has shown T_m parameter 13.8°C, while T_m for hybrid $d(T_{Me})_7T$ random (5)/4 gave similar value. Interestingly, T_m for hybrid 1/4 was 38°C, while that for 2/4 was <2°C. These measurements clearly speak for development of methods leading to P-stereoregular analogues of DNA, considered as hybridization probes and modulators of gene expression[2], and prompted us to further studies on development of stereoselective methods of synthesis of all-phosphorothioate analogues of DNA (oligo-S)[3].

The synthesis of 5' MMT-thymidine-3'-O-[O-(4-nitrophenyl)-S-(2-nitrobenzyl)phosphorothioate] and separation of its diastereomers allowed their reaction with 3'-acetylthymidine. t-BuMgCl has been used as catalyst for this process[4]. Purification of the product, after treatment with thiophenol/triethylamine (S-deprotection) and usual work-up, gave R_P- or S_P-$T_{PS}T$ (6), respectively. On the similar way (R_P,R_P)-$T_{PS}T_{PS}T$ and (S_P,S_P)-$T_{PS}T_{PS}T$ (7), were obtained in moderate yield; their identification was performed by means of HPLC and coinjection of resulting material with genuine samples of diastereomers of $T_{PS}T_{PS}T$ of known absolute configuration at phosphorus, obtained by literature methods[5].

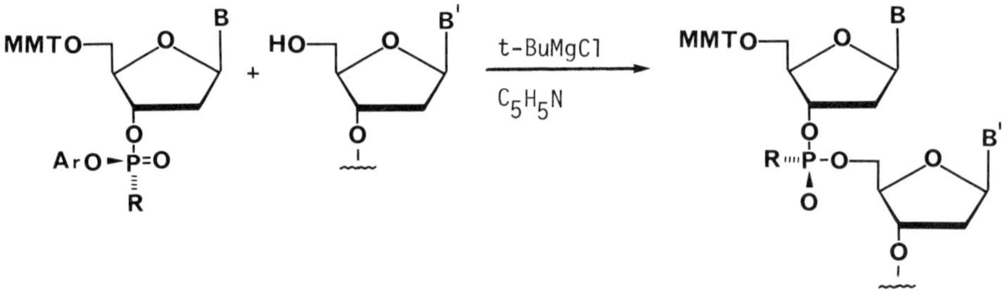

R = CH_3-, 2-NO_2-C_6H_4-CH_2S-
Ar= 4-NO_2-C_6H_4-

Further studies on the synthesis of stereoregular P-chiral oligo(nucleoside methanephosphonates) and oligo(nucleoside phosphorothioates) are in progress.

References

1.Lesnikowski, Z.J., Jaworska, M., Stec, W.J. (1988) Nucleic Acids Res., 24, 11675-11689.
2.Zon, G. (1988) Pharm.Res., 5, 539-549.
3.Matsukura, M., Shinozuka, K., Zon, G., Mitsuya, H., Reitz, M., Cohen, J.S., Broder, S. (1987) Proc.Natl.Acad.Sci., USA, 84, 7706-7710.
4.Lesnikowski, Z.J., Jaworska, M. (1989) Tetrahedron Lett., submitted for publication.
5.Stec, J.W., Zon, G. Egan, W. Stec, B.(1984) J.Am.Chem.Soc., 106, 6077-6079

Symposium Series No. 21 1989 Nucleic Acids Research

Improved synthesis of oligodeoxyribonucleotides

Klaus-Peter Stengele and Wolfgang Pfleiderer

Fakultät für Chemie, Universität Konstanz, Postfach 5560, D-7750 Konstanz, FRG

ABSTRACT

An improved synthesis of oligodeoxyribonucleotides was achieved with NPE- and NPEOC-protected phosphoramidites leading to easier and time-saving purification and isolation of the gene fragments.

The phosphoramidite approach revolutionized oligonucleotide synthesis from a chemical point of view and allowed the development of automated processes for the built-up of DNA-fragments

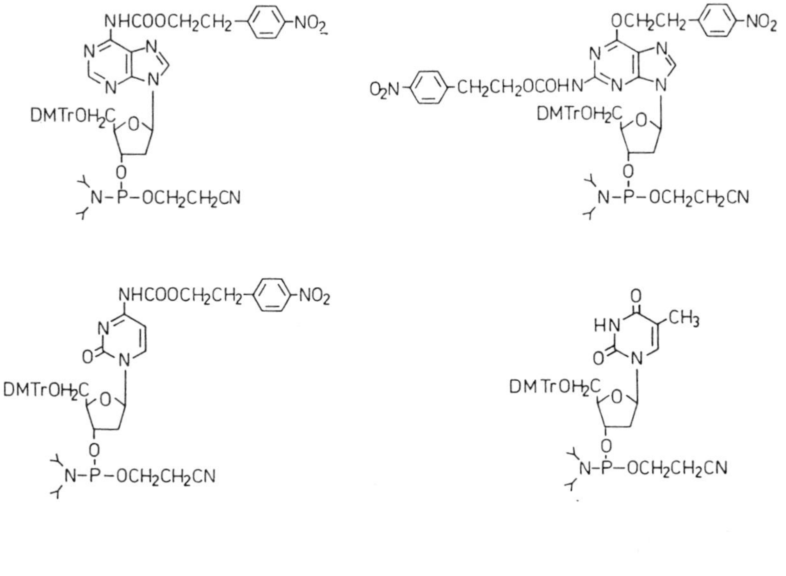

(1-3). A great number of variations of the original method have been published and recommended including mainly changes of the blocking group patterns at the aglycone, sugar and phosphite moiety of the monomeric building blocks to get higher yields and greater purity of the products. Since chemical reactions can always be improved we developed a new set of ß-cyanoethylphosphor-N-diisopropyl-amidites containing the p-nitrophenylethyl (NPE) and p-nitrophenylethoxycarbonyl (NPEOC) groups (4) for amide and amino protection of the pyrimidine and purine bases.

The solid support material consisting of CPG-500 A and CPG-1400 A particles was also modified with 1,6-bis-methylaminohexane as an aliphatic spacer to increase the stability of the linkage between the oligonucleotide chain and the solid matrix.

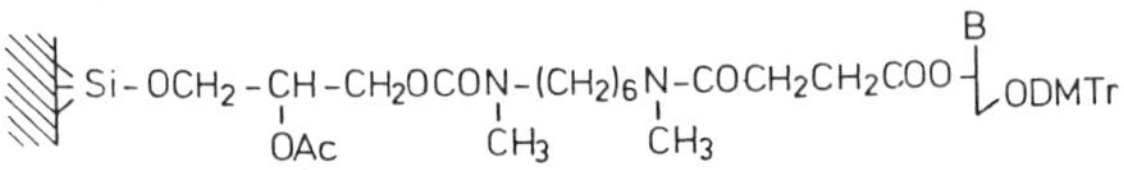

The chain elongation cycle was slightly modified and gave condensation yields ranging usually between 99 and 99.5 %.

The big advantage of the new strategy became obvious during the deprotection procedure and the simplified, time-saving isolation of the synthesized oligodeoxyribonucleotide. After completing the final cycle fully automated deblocking was done first by the trityl-off and washing step and then by 6 h treatment with 0.5 M DBU in CH_3CN to remove all the cyanoethyl, NPE- and NPEOC-groups. Final washing gave the fully deprotected oligonucleotide still attached to the solid support. Cleavage was then achieved by ammonia treatment at room temp. for 4 h and purification and direct isolation by ion-exchange HPLC-chromatography on a Nucleogen 60-7 DEAE column (5). The elution profile revealed very small amounts of impurities of the shorter sequences but a homogeneous high peak of the anticipated oligonucleotide. Various 40-mers have been synthesized by the standard method using the commercially available phosphoramidites as well as the NPE/NPEOC-protected analogs to demonstrate by comparative studies the superiority of the latter components what the purification and the isolation of the oligonucleotide is concerned.

The improvements of our new modified approach to achieve automated oligodeoxyribonucleotide syntheses in high yields will be discussed in great detail.

REFERENCES

1. Dorman, M.A., Noble, S.A., McBride, L.J. and Caruthers, M.H. (1984) Tetrahedron 40, 95.
2. Köster, H., Biernat, J., McManus, J., Wolter, A., Stumpe, A., Narang, CH.K., and Sinha, N.D. (1984) Tetrahedron 40, 103.
3. Gassen, H.G. and Lang, A. (1982) Chemical and Enzymatic Synthesis of Gene Fragments, Verlag Chemie, Weinheim.
4. Himmelsbach, F., Schulz, B.S., Trichtinger, T., Charubala, R. and Pfleiderer, W. (1984) Tetrahedron 40, 59.
5. Macherey-Nagel, D-5160 Düren / West Germany.

General, regiocontrolled synthesis of unsymmetrically linked dinucleoside adenosine 2′,3′-bis(phosphate)s

Yoshihiro Hayakawa, Masaaki Hirose[1], and Ryoji Noyori[1]

The Chemical Instrument Center and [1]Department of Chemistry, Nagoya University, Chikusa, Nagoya 464-01, Japan

ABSTRACT

Adenyl ribonucleotides having unsymmetrical 2'-5',3'-5' phosphodiester linkage have been prepared in a general, regiodefined manner.

INTRODUCTION

2'-5',3'-5'-Linked adenyl oligoribonucleotides (branch-type RNAs) firstly discovered in an intrinsic part of eukaryotic RNA splicing[1] are conceived to have an important relation to enzymes, i.e. ribozymes, catalyzing self-splicing reaction.[2] Immunochemical studies on 2-5A have implied that the analogs having a nucleotide branch at the 3' position are attractive compounds for chemotherapeutical use. This paper describes a convenient, general synthesis of adenyl branch-type oligonucleotides using allyl and allyloxycarbonyl (AOC) groups as protectors of internucleotide linkage and nucleobases, respectively.[3]

RESULTS AND DISCUSSION

Condensation of the $\underline{N}^6$-allyloxycarbonyl-3'-O-tert-butyldimethylsilyl-5'-O-p-methoxytrityladenosine and the 5'-phosphoramidite **1** in the presence of 1H-tetrazole and 4-dimethylaminopyridine (DMAP) followed by oxidation with tert-butyl hydroperoxide (TBHP) gave the 2'-5'-linked phosphotriester **5** in 99% yield. Selective deallylation of **5** with an excess of NaI in acetone and subsequent desilylation with tetrabutylammonium fluoride furnished the 3'-O-free derivative **6**. No undesired migration of the 2'-phosphate moiety to the 3'-position took place during these transformations. The tetrazole/DMAP-assisted coupling of **6** and the 5'-phosphoramidite **2** followed by TBHP oxidation afforded the branch-type RNA **7** in 78% yield, which was then deprotected by exposure to dichloroacetic acid followed by $Pd[P(C_6H_5)_3]_4$ (5 mol%/allyl), $P(C_6H_5)_3$ (3 mol%/allyl), and a 1:1:1 mixture of H_2O, CO_2, and $\underline{n}$-$C_4H_9NH_2$ in THF, to give the target compound **8** in 80% yield. This synthetic approach is general and practical. Thus unsymmetrically linked dinucleoside adenosine 2',3'-bis-(phosphate)s such as **9—12** were also prepared in a similar manner using **1—4** as the 5'-phosphoramidite units. In all cases, the target products precipitated out during the reaction and the pure samples could be isolated by simple recrystallization from ethanol.

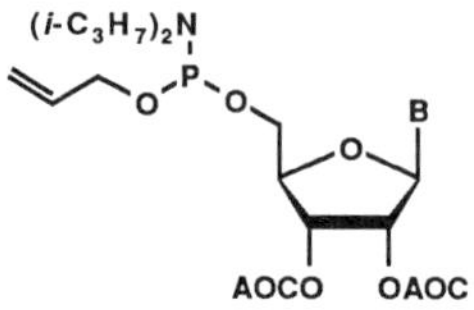

1, B = Ur
2, B = Ad^{AOC}
3, B = Cy^{AOC}
4, B = Gu^{AOC}

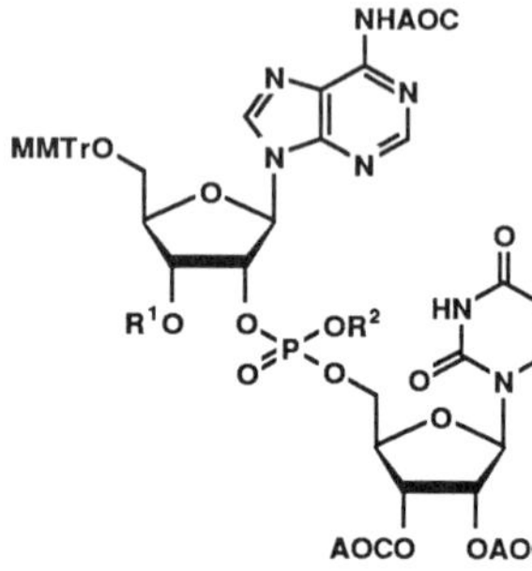

5, R^1 = TBDMS; R^2 = allyl
6, R^1 = H; R^2 = Na

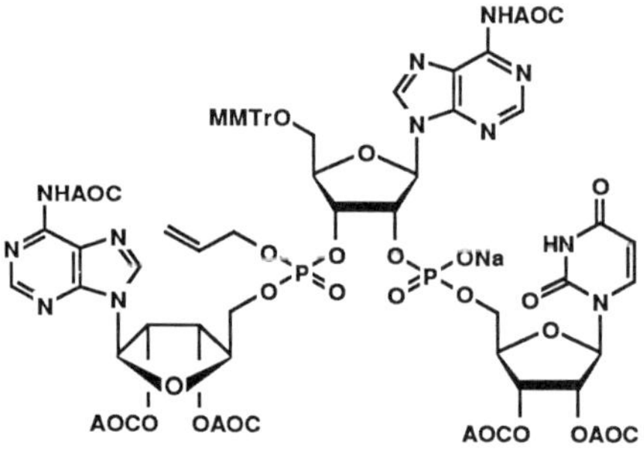

7

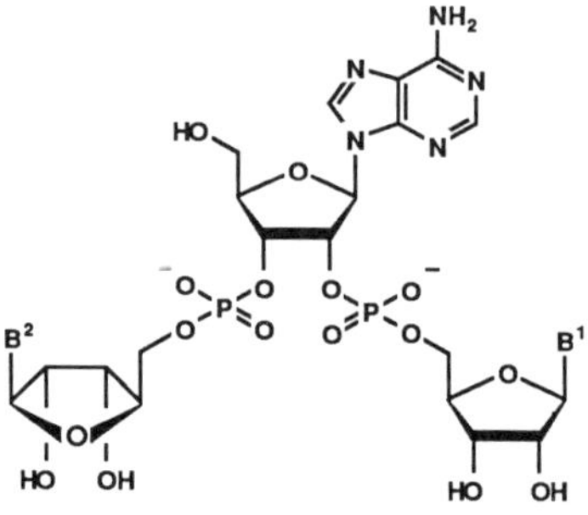

8, B^1 = Ur; B^2 = Ad
9, B^1 = Ad; B^2 = Cy
10, B^1 = Ad; B^2 = Gu
11, B^1 = Ad; B^2 = Ur
12, B^1 = Cy; B^2 = Ad

REFERENCES

1. Wallace,J.C. and Edmonds,M. (1983) Proc.Natl.Acad.Sci.U.S.A., **80**, 950.
2. Cech,T.R. and Bass,B.L. (1986) Ann.Rev.Biochem., **55**, 599.
3. (a) Hayakawa,Y., Uchiyama,M., Kato,H. and Noyori,R. (1985) Tetrahedron Lett., **26**, 6505. (b) Hayakawa,Y., Kato,H., Uchiyama,M., Kajino,H. and Noyori,R. (1986) J.Org.Chem., **51**, 2400.

Synthesis and properties of RNA with catalytic activity

Osamu Odai, Takeshi Sakata, Masaya Orita, Hidekazu Hiroaki, Seiichi Uesugi, and Toshiki Tanaka[1]

Faculty of Pharmaceutical Sciences, Osaka University and [1]Protein Engineering Research Institute, Suita, Osaka 565, Japan

ABSTRACT

Ribooligonucleotides, which form the self-cleavage domain of a satellite RNA of tobacco ringspot virus, were synthesized by the solid-phase phosphoramidite method using o-nitrobenzyl groups for 2'-hydroxyl protection. The oligomers were obtained in quantities sufficient for NMR measurement. Specific cleavage at an expected site was observed when the three RNA fragments were mixed in the presence of magnesium ions.

INTRODUCTION

Relatively long RNA oligonucleotides can be prepared by chemical synthesis using the solid-phase phosphoramidite (1) or H-phosphonate (2) methods or by enzymatic synthesis using viral RNA polymerases. However it is still difficult to obtain RNA oligomers in quantity. Recently we succeeded to synthesize RNA oligomers with chain lengths of 6∿9 nucleotides by the solid-phase phosphoramidite method in sufficient amounts for NMR study. In this paper, we report synthesis and properties of RNA oligomers with chain lengths of 11∿16 nucleotides which self-cleaves at a specific site when assembled in the presence of Mg^{2+}.

RESULTS AND DISCUSSION

Three oligonucleotides, CCUGUCACCGC (1), GGACGAAACAGG (2) and GCGGUCUGAUGAGUCC (3), were synthesized by the solid-phase phosphoramidite method using o-nitrobenzyl groups for 2'-OH protection and cyanoethyl groups for phosphorus protection. These oligomers contain base sequences of the self-cleavage domain of a satellite RNA of tobacco ringspot virus(3). The reaction cycle for the solid-phase synthesis was as reported previously (4). After partial deprotection with concentrated ammonia, the monomthoxytritylated oligomer was isolated by reversed phase column chromatography. After detritylation with 80% aqueous acetic acid, the o-nitrobenzyl groups were removed by UV irradiation at pH 3.5. Gel filtration and reversed phase column chromatography gave pure oligomers (1∿1.5 μmoles) in a yield of 10∿25%. The amounts of the oligomers thus obtained are sufficient for high-field NMR measurement to examine the detailed conformation. When these oligomers (1∿3) were mixed in the presence of Mg^{2+}, 1 was cleaved at an expected site (between 6C and 7A). When the 6C of 1 was replaced with 2'-O-methylcytidine, the corresponding oligonucleotide complex gave no cleavage product. Effects of temperature and cationic species on the cleavage reaction were also examined. Conformation of the complex is now under investigation.

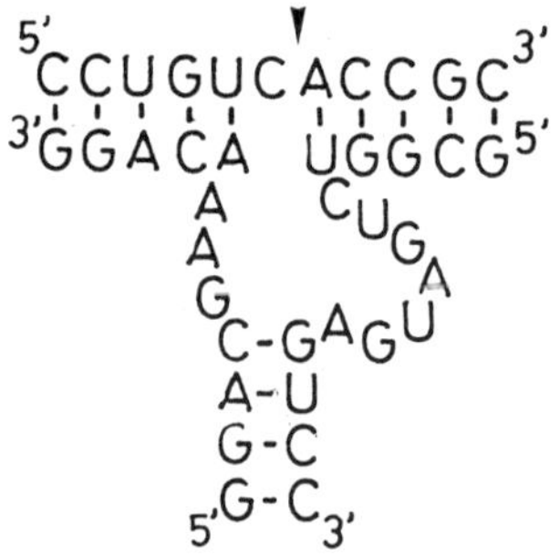

REFERENCES

1. Usman,N., Ogilvie,K.K., Jiang, M.-Y. and Cedergren,R.J. (1987) J. Am. Chem. Soc. 109, 7845-7854.
2. Tanaka,T., Tamatsukuri,S. and Ikehara,M. (1987) Nucleic Acids Res. 15, 7235-7248.
3. Prody,G.A., Bakos,J.T., Buzayan,J.M., Schneider,I.R. and Bruening,G. (1986) Science 231, 1577-1580.
4. Tanaka,T., Tamatsukuri,S. and Ikehara,M. (1986) Nucleic Acids Res. 14, 6265-6279.

Design of RNA enzymes for sequence-dependent cleavage of RNA

Makoto Koizumi, Yoji Hayase, Shigenori Iwai, Hiroyuki Kamiya, Hideo Inoue, and Eiko Ohtsuka

Faculty of Pharmaceutical Sciences, Hokkaido University, Sapporo 060, Japan

ABSTRACT

We have designed synthetic ribozymes which can cleave specific sites of RNA. An undecamer, 5S rRNA and transcripts using T7 RNA polymerase were hydrolyzed at the X position of RNA containing $(N)_nUX(N)_n$ (N=A, G, C and U; X=A, C and U) by these ribozymes.

It is known that RNA processing reactions are catalyzed by RNA enzymes. One of these reactions, the self-cleavage reaction, which has been found in satellite RNA of viruses, infectious RNA in plants or transcripts of satellite DNA of the newt,[1] is believed to be an essential step during the replication cycle of these RNAs. This reaction requires divalent ions and neutral or alkaline pH. The cleavage products contain 2', 3'-cyclic phosphodiester and 5'-hydroxy termini. The self-cleavage domain containing 3 stems and 13 conserved nucleotides, termed 'hammerhead', has been postulated. Some 'hammerhead' domains have been prepared using T7 RNA polymerase, and were proved to be important for the self-cleavage reaction.[2]

We have also constructed a series of several self-cleaving RNA duplexes using chemically synthesized 21mers.[3] On the basis of the results obtained from these experiments, we have designed RNA enzymes for sequence dependent cleavage of RNA.[4] Recently, Haseloff and Gerlach reported that RNA enzymes could cleave chloramphenicol acetyl transferase gene transcripts, but these ribozymes targeted against the substrate containing sequence only GUC adjacent to the site for cleavage.[5]

The design of ribozymes so that the 3'-site of UX (X=A, C

and U) in various RNAs can be cleaved efficiently as shown in figure, is based on previous observations.[3] The other nucleotides represented by N in the substrate form Watson-Crick base pairs with the nucleotides of the ribozyme. The other sequences in the ribozyme comprise the sequences of newt satellite DNA transcripts[1] containing 11 highly conserved nucleotides.

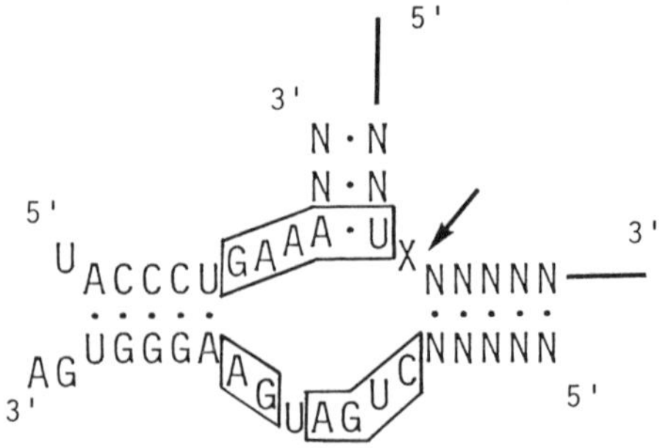

An undecamer, pCAGCUAAGUAU, having a sequence similar to the 5'-splice sequence was selected as the substrate and A6 as a cleavage site. The ribozyme, which can cleave this substrate, consisted of a 19mer and a 15mer. This complex was hydrolyzed at the expected position. A Hanes-Woolf plot, which was used to determinate the kinetic parameters, showed that this ribozyme cleaved the substrate catalytically. The Km and Vmax were found to be 0.53µM and 0.015$\mu Mmin^{-1}$, respectively. These results are similar to those of Uhlenbeck's system.[2] Ribozymes which delete a number of hydrogen bonds with either the 5'- or 3'-end of the substrate, reacted more slowly. On the other hand, a complex containing a mismatch C-C in the synthetic ribozyme, was not hydrolyzed. Thus, hybridization of the ribozyme to the substrate affects catalytic reactions, and the base-pairing stability of the duplexes is important for the rate of cleavage of the substrate.

We have also applied this technique to naturally occuring 5S rRNA from Desulfovibrio vulgaris MK and the RNA transcripts obtained using synthetic DNA and T7 RNA polymerase.

REFERENCES

1. Epstein, L. M. and Gall, J. G. (1987) Cell 48, 535-543.
2. Uhlenbeck, O. C. (1987) Nature 328, 596-600.
3. Koizumi, M., Iwai, S. and Ohtsuka, E. (1988) FEBS Lett. 228, 288-230.
4. Koizumi, M., Iwai, S. and Ohtsuka, E. (1988) FEBS Lett. 239, 285-288.
5. Haseloff, J. and Gerlach, W. L. (1988) Nature 334, 585-591.

Synthesis and biologic activity of nucleoside analogs

John A.Montgomery

Southern Research Institute, 2000 Ninth Avenue, South P.O. Box 55305 Birmingham, AL 35255-5305, USA

There are at least 85 enzymatic reactions involved in the *de novo* synthesis of purine and pyrimidine nucleotides, in their interconversions and in their polymerization to nucleic acids, and in the so-called salvage pathways. Of these 85 enzymes, approximately 14 are known to be inhibited by metabolite analogs or anabolites thereof. These inhibitions are thought to be responsible for, or at least to contribute to, the biologic activity of these compounds. Incorporation into nucleic acids, blocking their function, appears to be responsible for the activity of some of the antimetabolites, although enzymatic blocks probably also play a role. The differences in the various antimetabolites are evident from their metabolism, the lack of cross-resistance between them, and their therapeutic synergism.

The importance of metabolism to the action of purine-pyrimidine antimetabolites has long been appreciated, but only more recently have we come to realize that we can, with limitations, control drug metabolism by structural alterations that do not destroy desired activity. In many cases, more active drugs have resulted. In others, we have learned more about the enzymes that metabolize purines, pyrimidines, and their nucleosides and nucleotides.

Certain structural features of the naturally occurring purine and pyrimidine nucleotides determine the enzymes that metabolize them, and an analysis of this metabolic picture allows the classification of antimetabolites and predictions about their metabolism, although this classification has sometimes been deceptive and made correctly only in retrospect.

Using this information, we have designed a variety of nucleoside analogs with useful biologic activity including nucleosides of aza- and deazapurines and haloadenines, and carbocyclic analogs. Recently we have explored the synthesis and biologic activity of a number of 4′-thionucleosides.

The chemistry leading to a variety of structural modifications of nucleosides, and the effects of these modifications on biologic activity, as reflected in altered metabolism and mechanism of action, will be discussed.

Modification of 5-methylcytosine in DNA and its relation to mutagenesis

Kazuo Negishi[1], Yoshifumi Harada, Gen-ichi Atsumi, Tadayoshi Bessho, Tomoe Negishi, and Hikoya Hayatsu

Faculty of Pharmaceutical Sciences and [1]Gene Research Center, Okayama University, Tsushima, Okayama 700, Japan

5-Methylcytosine is a minor base of DNA in various organisms. In mammalian cells, the 5-methylcytosine content in DNA is about 5 % of cytosine. 5-Methylcytosine has been found to be readily oxidized with potassium permanganate[1]. The rate of the reaction is similar to that of the reaction of thymine with permanganate. In contrast, cytosine is oxidized more slowly and purines are much less reactive to permanganate. Products from the reaction of 5-methyldeoxycytidine with permanganate have been fractionated with HPLC and the identification of these products are underway.

This high susceptibility of 5-methylcytosine to oxidation with permanganate suggests the possibility that 5-methylcytosine in DNA is a mutational hot spot toward the action of oxidative agents. Although thymine bases are thought to be most susceptible to modification by oxidation, the major product, thymine glycol, was shown to be non-mutagenic in the phage M13mp2 system[2]. To investigate the correlation between modification of 5-methycytosine in DNA and mutagenesis, we have constructed a new assay system using a mutant strain of M13mp2. The mutant used is M13mp2-87C, which is one of the α-complementation deficient strains produced by the action of a nucleoside mutagen, N^4-aminocytidine[3]. This mutation from T to C at position +87 has created a new *Hpa* II site (5´-C↑CGG-3´) as shown in Figure 1. The base which has been mutated to C can be methylated *in vitro* by the action of Hpa II methylase. When the DNA produced is subjected to an oxidative reaction and subsequent transfection to the bacteria, an efficient production of revertants may be expected to occur due to the potentially high reactivity of this 5-methylcytosine residue: the DNA that has been methylated is efficiently reverted, while the DNA before the *Hpa* II methylase treatment is not. Using these monitoring techniques, we should be able to test our hyposesis that 5-methylcytosine can be a hot spot.

5´-T CGT GAC TGG G--3´ DNA sequence of a wild type M13mp2
+80 +87 (Numbers indicate the distance from the nucleotide where transcription starts.)

Mutagenesis by N^4-aminocytidine

Hpa II site

3´-A GCA CTG GCC C--5´
5´-T CGT GAC CGG G--3´ A viral strand; a mutant, M13mp2-87C

In vitro methylation with *Hpa* II methylase

3´-A GCA CTG G ^{m}CC C--5´ (^{m}C; 5-methlcytosine)
5´-T CGT GAC ^{m}C GG G--3´

Treatment with an oxidative agent
Transfection to bacteria

5´-T CGT GAC TGG G--3´ Revertant

Figure 1. Scheme for assaying mutability of a methylated site in DNA.

REFERENCES

1. Fritzsche, E., Hayatsu, H., Igloi, G. L., Iida, S. and Kössel, H. (1987) Nucleic Acids Res., 15, 5517-5528.
2. Hayes, R. C., Petrullo, L. A., Huang, H., Wallace, S. S and LeClerc, J. E. (1988) J. Mol. Biol., 201, 239-246.
3. Bessho, T., Matsumoto, K., Nomura, A., Hayatsu, H. and Negishi, K. (1989) J. Mol. Biol., 205, 659-664.

Some properties of bovine mitochondrial serine tRNA gene transcript synthesized with T7 RNA polymerase

Takuya Ueda and Kimitsuna Watanabe

Department of Life Chemistry, The Graduate School at Nagatsuta, Tokyo Institute of Technology, Nagatsuta, Yokohama 227, Japan

Abstract

Bovine mitochondrial serine tRNA(AGY) gene transcript was synthesized in vitro with T7 RNA polymerase, and it was capable of being aminoacylated with mitochondrial serine tRNA synthetase. The melting profiles of the transcript was similar to those of native serine tRNA(AGY), suggesting that the higher-order structure of the transcript does not much differ from that of native serine tRNA(AGY). Several transcripts with base-substitution were also constructed and their aminoacylation capacity was investigated.

Introduction

Mammalian mitochondrial serine tRNA, decoding AGY codons, [$tRNA^{Ser}$(AGY)] lacks the D stem and loop structure (1), which has been considered to be important for recognition by aminoacyl-tRNA synthetase. The fact that this unusual $tRNA^{Ser}$(AGY) functions in the organella, provides us a new clue for building a model for a primitive tRNA whose structure might have been more simplified. In order to study the structure-function relationship of mitochondrial $tRNA^{Ser}$(AGY), the in vitro transcription system with T7 RNA polymerase was employed for synthesizing the tRNA and its mutants as a substrate of aminoacylation. Their aminoacylation efficiencies as well as their structures were investigated.

Experiments

The DNA fragment consisting of the T7 promoter and bovine mitochondrial $tRNA^{Ser}$(AGY) gene was inserted into the multi-cloning site of pUC19. The tRNA region in the plasmid was transcribed by the method developed by Sampson and Uhlenbeck (2) with a slight modification, with T7 RNA polymerase purified from its overproducing strain (3) which was kindly provided by Prof. Studier. The synthesized tRNA was purified by electrophoresis. 0.5-1.0 A260 units of tRNA could be obtained, which were enough for enzyme assay. It is calculated that during the reaction the template DNA was utilized as much as approximately 100 times. The aminoacylation efficiencies and melting profiles were measured according to Yokogawa et al. (4).

Results and discussion

The synthetic $tRNA^{Ser}$(AGY) transcript was able to be aminoacylated by S-100 fraction prepared from bovine heart mitochondria in the similar manner as native $tRNA^{Ser}$(AGY). This result indicates that the t^6A nucleoside adjacent to 3' end of the anticodon, which is a single modified nucleoside in native $tRNA^{Ser}$(AGY), does not affect aminoacylation activity fundamentally.

The structure of the synthetic tRNA transcript was analyzed by measuring its melting profiles in comparison with native $tRNA^{Ser}$(AGY). The transition points of the transcript were 42°C and 58°C at 0 mM and 10 mM Mg^{2+}, respectively. Since these values are similar to those of the native tRNA, the structure of the transcript may be considered not to be quite different from that of naive tRNA. These results strongly suggested that the synthetic transcript can be utilized for the study of the structure-function relationship of tRNA as its substitute, because some tRNAs like those of mitochondria are very difficult to be purified in amounts sufficient for biochemical study.

The A2-U58 base pair, which is a single conserved base pair among nine $tRNA^{Ser}$(AGY) species of animal mitochondria, was changed to C2-G58 pair. This substitution did not affect aminoacylation capacity of the modified tRNA. Since this base-pair may correspond to a part of the identity determinant region for E. coli serine tRNA, which Normanly et. al.(5) pointed out, it can be speculated that a different recognition mechanism of aminoacyl-tRNA synthetase may exist in mammalian mitochondria. We are also analyzing the derivatives in which mutations have been introduced in other parts, such as anticodon region, discriminator base and other regions.

Reference

1. Aracri, P. and Brownlee, G. G. (1980) Nucleic Acids Res., 8, 5207-5212.
2. Sampson, J. R. and Uhlenbeck, O. C. (1988) Proc. Natl. Acad. Sci. USA, 85, 1033-1037.
3. Davanloo, P., Rosenberg, A. H., Dunn, J. J. and Studier, F. W. (1984) Proc. Natl. Acad. Sci.USA 81, 2035-2039.
4. Yokgawa, T., Kumazawa, Y., Miura, K. and Watanabe, K. (1989) Nucleic Acids Res. 17, 2623-2638
5. Normanly, J., Ogden, R. C., Horvath, S. J. and Abelson, J. (1986) Nature 321, 213-219.

Differential inhibition of HIV-reverse transcriptase and various DNA and RNA polymerases by some catechin derivatives

Hideo Nakane and Katsuhiko Ono

Laboratory of Viral Oncology, Aichi Cancer Center Research Institute, Chikusa-ku, Nagoya 464, Japan

ABSTRACT

The two components of *Camellia sinensis* (green tea) (i.e., (-)-epigallocatechin gallate and (-)-epicatechin gallate were found to differ-erentially inhibit the activities of reverse transcriptase and cellular DNA and RNA polymerases. The strongest inhibition by these compounds was observed with reverse transcriptase. DNA polymerases α and β were also strongly inhibited. The mode of the inhibition of reverse transcriptase and other DNA polymerases was competitive with respect to the template·primer, whereas the mode the inhibition of RNA polymerase was competitive with respect to the nucleotide substrate.

INTRODUCTION

Since acquired immune deficiency syndrome (AIDS) was found to be caused by a retrovirus, designated as human immunodeficiency virus (HIV), HIV-associated reverse transcriptase (RT) has been considered to be one of the appropriate targets for chemotherapeutic approaches towards AIDS. However, most of the antivirals which inhibit RT also inhibit cellular DNA polymerases. This explains at least a part of the side effects observed upon administration of these anti-HIV agents to AIDS patients. Typical examples are suramin and HPA23 which inhibit both RT and cellular DNA polymerases,[1,2] and both compounds have revealed various side effects among the patients.

To find more selective but less toxic compound(s), we have been looking for natural product(s) inhibitory to RT.[3] We found that (-)-epicatechin gallate (CG) and (-)-epigallocatechin gallate (GG), the components of *Camellia sinensis*, were strong inhibitors of HIV-RT. In this paper, the inhibitory effects of CG and GG on the activities of various nucleotide polymerizing enzymes are described.

MATERIALS AND METHODS

CG and GG were the products of Mitsui Norin Co. Ltd., Japan.

HIV-RT was purified from *E. coli* harboring an expression plasmid for the precise coding sequence of the enzyme (a kind gift from Dr. S.H. Wilson, NIH, USA). RLV-RT, DNA polymerases α, β and γ, and terminal deoxynucleotidyltransferase (TdT) were purified from Rauscher murine leukemia virus (RLV), cultured human KB III cells, and calf thymus, respectively.[1] Assay for DNA polymerase activities were carried out under the reaction conditions described previously.[1] Inhibitory potentials of the compounds were expressed as inhibition constants (Ki's) obtained by the kinetic analysis.

RESULTS AND DISCUSSION

1) Both CG and GG strongly inhibited HIV-RT (50% inhibition dose, IC_{50}, of both compounds were about 0.02 μg/ml). On the other hand, (-)-epicatechin, (-)-epigallocatechin and gallic acid, constituents of CG and/or GG, were not inhibitory at concentrations up to 1 μg/ml.
2) DNA polymerases α and β, and RLV-RT were also inhibited by any of the compounds (IC_{50}: < 0.2 μg/ml).
3) DNA polymerase γ was moderately inhibited by CG and GG, whereas TdT was not inhibited by these compounds.
4) The mode of the inhibition of RT and other DNA polymerases by these compounds was competitive with respect to the template·primer except for HIV-RT (mixed type).
5) The mode of the inhibition of RNA polymerase was noncompetitive with respect to poly(dC) template, and competitive to GTP substrate.

All the results obtained in the present study indicate that both CG and GG are very strong inhibitors particularly for HIV-RT. It is possible that these compounds exert selective inhibition against HIV-RT, when appropriate concentrations of the compounds are chosen.

REFERENCES

1. Ono, K., Nakane, H. and Fukushima, M. (1988) Eur. J. Biochem., 172, 349-353.
2. Ono, K., Nakane, H., Barré-Sinoussi, F. and Chermann, J.-C. (1988) Eur. J. Biochem., 176, 305-310.
3. Ono, K., Nakane, H., Meng, Z.-M., Ose, Y., Sakai, Y. and Mizuno, M. (1989) Chem. Pharm. Bull., in press.

Mg^{2+} effects on a circle opening reaction of the self-splicing intervening sequence from *Tetrahymena thermophila*

Naoki Sugimoto*, Muneo Sasaki, Ryszard Kierzek[1], and Douglas H.Turner[2]

Department of Chemistry, Faculty of Science, Konan University, Kobe 658, Japan, [1]Institute of Bioorganic Chemistry, Polish Academy of Science, Poznan, Poland, and [2]Department of Chemistry, University of Rochester, Rochester, NY 14627, USA

ABSTRACT

The Mg^{2+} contibution to the reaction of circular intervening sequence (CIVS) from rRNA precursor of Tetrahymena thermophila with a dinucleotide CU has been investigated. The results indicated that the circle opening of CIVS may involve binding of a weakly held Mg^{2+} ion.

INTRODUCTION

Self-splicing RNA may represent a molecular fossil since it can both code information and have catalytic activity such as a ribozyme. The first RNA known to facilitate cleavage and ligation reactions in RNA is the intervening sequence (IVS) from the rRNA precursor of Tetrahymena thermophila (1-3). This RNA mediates a self-excision coupled to an exon splicing reaction and subsequently undergoes a self-circularization. The circular product (CIVS) can be opened by addition of oligonucleotides, and this new linearized product undergoes itself circularization in a reversible manner (4). The detailed mechanism of the reaction is the subject of continued investigation, because its reaction provides a convenient model system for studying the mechanism of many IVS reactions. The recent study has indicated that binding of oligonucleotide substrates is very large and that a weakly bound Mg^{2+} ion may be required for reaction (5).

In this paper, we report the effect of Mg^{2+} on the reaction mechanism for the linearization of CIVS with a dinucleotide CU. Our kinetic analysis and its results suggest that a specific binding like a formation of an inner-sphere complex may be necessary in the IVS reaction.

RESULTS AND DISCUSSION

The CIVS opened specifically at G_{414}-A_{16} phosphodiester linkage by attack of CU. The extent of reaction on time was analyzed with a program of the single exponential fit to give an observed rate constant k_{obs}. At 30 °C and 10 mM Mg^{2+}, k_{obs} was essentially saturated at 1 mM CU. The almost saturated values of k_{obs} at 1 mM CU depended on the initial concentration $[Mg^{2+}]_0$, and the results indicated the binding constant K_a of Mg^{2+} with CIVS was 513 M^{-1} at the condition. Surprisingly, the value of K_a is ten to a hundred fold smaller than that commonly observed for Mg^{2+} binding to RNA (5). The two possible reasons can be considered: 1) Because the mechanism of Mg^{2+} binding was treated in the kinetically saturated CU concentration, the unusually small value of K_a is only in an extremely simple mechanism of the reaction. K_a should increase if the reaction could be treated by a general kinetic equation considering both binding steps by CU and Mg^{2+}. 2) If the very small value of the binding constant is observed in a general kinetic treatment, such small affinity of Mg^{2+} may reflect a specific binding to CIVS.

The kinetic treatment of the general mechanism in which Mg^{2+} bound CIVS after the CU substrate showed that the binding constant of Mg^{2+} to CIVS was only 4% larger than that in the simple kinetic treatment. Therefore, the mechanism of circular opening by CU may involve a weak binding of Mg^{2+} in an inner sphere complex involving the ribose of 3'U of the substrate.

REFERENCES

1. Kruger, K., Grabowski, P. J., Zaug, A. J., Sands, J., Gottschling, D. E., and Cech, T. R. (1982) Cell (Cambridge, Mass.) **31**, 147-157.
2. Zaug, A. J., Grobowski, P. J., and Cech, T. R. (1983) Nature (London) **301**, 578-583.
3. Cech, T. R. and Bass, B. L. (1986) Ann. Rev. Biochem., **55**, 599-629.
4. Sullivan, F. X. and Cech, T. R. (1985) Cell (Cambridge, Mass) **42**, 639-648.
5. Sugimoto, N., Kierzek, R., and Turner, D. H. (1988) Biochemistry **27**, 6384-6392.

Synthesis of oligodeoxynucleoside phosphorodithioates

M.H.Caruthers, W.K.-D.Brill, A.Grandas, Y.-X.Ma, J.Nielsen, and J.-Y.Tang

Department of Chemistry and Biochemistry, University of Colorado, Boulder, CO 80309-0215, USA

Deoxyoligonucleotides containing phosphorodithioate internucleotide linkages (nucleoside-OPS_2O-nucleoside) are of considerable interest for many biochemical and biological applications because these compounds are isosteric and isoelectronic with normal phosphodiesters. These chemical and biochemical properties have prompted us to develop several approaches for synthesizing this internucleotide linkage (1-4).

The first step leading to **1**, the key intermediate for synthesizing phosphorodithioate linkages via two approaches (1,3), is condensation of a protected deoxynucleoside with bis(diisopropylamino)chlorophosphine. The resulting 3'-phosphorodiamidite is reacted without isolation with a 3'-protected deoxynucleoside to yield **1**. Synthesis of **4** can then proceed by (i) treatment with H_2S and tetrazole to yield **2**, (ii) oxidation with elementary sulfur, and (iii) alkylation with α,2,4-trichlorotoluene. Alternatively, **1** can be treated with (iv) 2,4-dichlorobenzylmercaptan and (v) elementary sulfur which again yields **4**. Conversion of **4** to the 3'-phosphoramidite using standard methodologies then produces a dinucleotide synthon useful for synthesizing oligonucleotides.

The most attractive approach utilizes deoxynucleoside 3'-phosphorothioamidites as synthons (2,4). These compounds are prepared by condensing a suitably protected deoxynucleoside with bis(dimethylamino)chlorophosphine and triethylamine to yield the diamidite (**5**) which is converted to the phosphorothioamidite (**6**) by addition (vi) of 2,4-dichlorobenzylmercaptan. After aqueous extraction and precipitation, these synthons can be stored as dry powders under inert gas atmospheres without decomposition. Synthesis of DNA begins (vii) by coupling (**6**) with a deoxynucleoside covalently linked to a silica support using tetrazole as an activator. This step is followed by (viii) sulfurization, (ix) capping with acetic anhydride, and (x) detritylation to yield **7**. Further repetitions of this cycle using either deoxynucleoside phosphorothioamidites or deoxynucleoside phosphoramidites (5) as synthons and tetrazole as an activator yield oligodeoxynucleotides having normal phosphate diester and phosphorodithioate diester linkages in any combination.

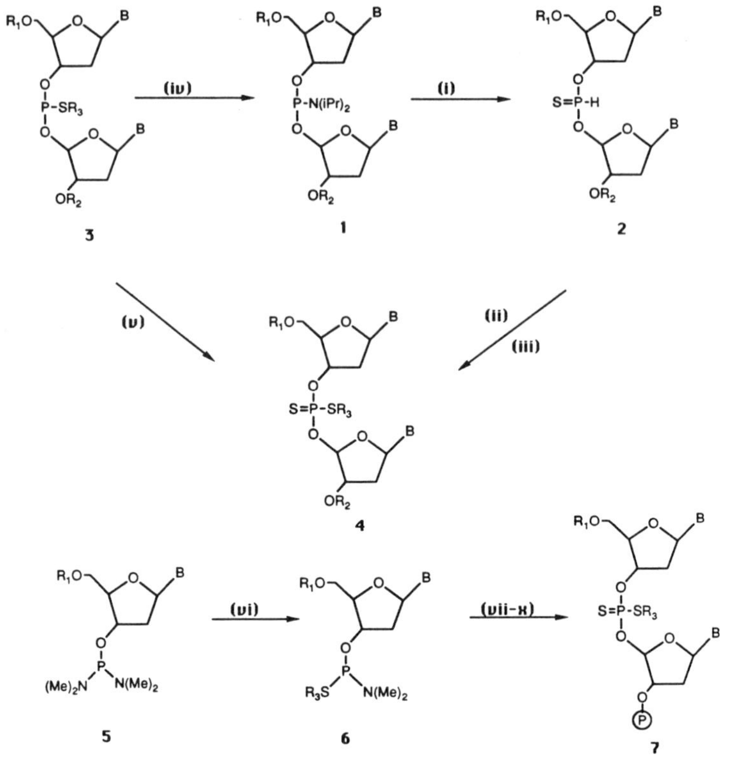

Synthesis of Phosphorodithioate DNA. Reactions (i-x) are outlined in the text. Abbreviations: R_1, dimethoxytrityl; R_2, acetyl or phenoxyacetyl; R_3, 2,4-dichlorobenzyl; Me, methyl; iPr, isopropyl; B, protected adenine, guanine, cytosine and thymine, Ⓟ , silica support.

ACKNOWLEDGEMENTS

This research was supported by research grants from NIH (GM21120 and GM25680).

REFERENCES

1. Nielsen, J., Brill, W. K.-D., and Caruthers, M. H. (1988) Tetrahedron Lett. 29, 2911-2914.
2. Brill, W. K.-D., Nielsen, J., and Caruthers, M. H. (1988) Tetrahedron Lett. 29, 5517-5520.
3. Grandas, A., Marshall, W., Nielsen, J. and Caruthers, M. H. (1989) Tetrahedron Lett. 30, 543-546.
4. Brill, W. K.-D., Tang, J.-Y., Ma, Y.-X, and Caruthers, M. H. (1989) J. Am. Chem. Soc. 111, 2321-2322.
5. Beaucage, S. L. and Caruthers, M. H. (1981) Tetrahedron Lett. 22, 1859-1862.

Synthesis of 1-(5′-*O*-benzoyl-3′-deoxy-β-D-glycero-pentofuran-2′-ulosyl)uracil and its α-anomer by [1,2]-hydride shift

Katsumaro Minamoto, Norio Fujiwara, Yuji Hamano, and Shoji Eguchi

Institute of Applied Organic Chemistry, Faculty of Engineering, Nagoya University, Furo-cho, Chikusa-ku, Nagoya, Japan

ABSTRACT

In contrast with direct tosylation of 5'-O-benzoyl- (1d) or 5'-O-pivaloyl-1-β-D-lyxofuranosyl-uracil (1e) with TsCl/pyridine, tosylation of the 2',3'-O-dibutylstannylene derivatives (4d,e) of these compounds proved to give the 3'-O-tosyl derivatives 2d,e selectively. Coversion of 2d as a model to 1-(5'-O-benzoyl-3'-deoxy-β-D-glycero-pentofuran-2'-ulosyl)uracil (5-β) by base-induced [1,2]-hydride shift was examined under various reaction conditions, and the α/β ratio of the product mixture (5-α,β) determined by ^{1}H NMR spectroscopy in each case. The BzOLi/DMF combination has proved to be most profitable for obtaining 5-β.

INTRODUCTION

Previously, we reported the first synthesis of 1-(3'-deoxy-β-D-glycero-pentofuran-2'-ulosyl)pyrimidines by selective elimination reactions of some derivatives of 1-β-D-lyxofuranosyl pyrimidines.[1-3] Recently, we have found that tosylation of some 5'-substituted derivatives (1b-e) of 1-β-D-lyxofuranosyl-uracil by a modified dibutylstannylation method[4] leads to predominant synthesis of the corresponding 3'-O-tosyl derivatives 2b-e, the yields being dependent upon bulkiness of the 5'-substituent.[5] This finding spurred us to examine the synthesis of the title compounds by [1,2]-hydride shift, using the most easily accessible 2d and several basic catalysts.

RESULTS

Compounds 1 were tosylated with 1.5-2.0 equiv TsCl in dry pyridine (1b-e) (Method A) or by treatment of the preformed 2',3'-O-dibutylstannylene intermediates 4a-e with 1.2 equiv TsCl in pyridine (Method B). In the cases of 1b,c, partial deacylation occurred

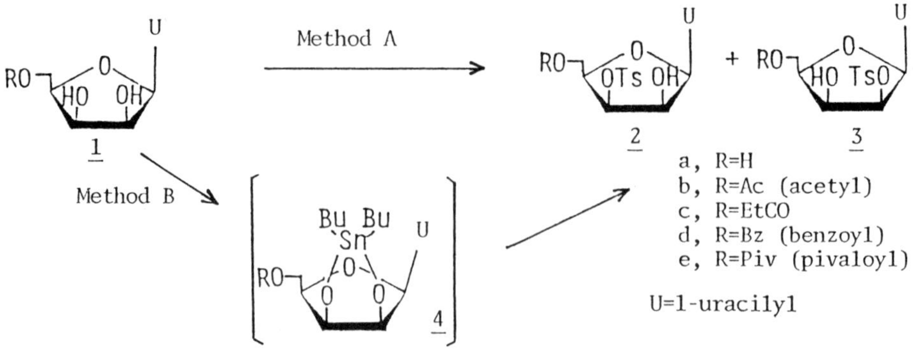

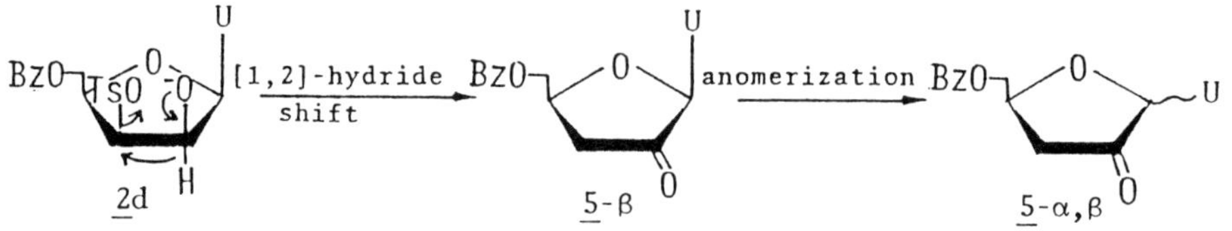

during the isolation procedure using silica gel chromatography, after tosylation, to give indefinite results.[6] However, Method B allowed highly selective synthesis of the 3'-O-tosyl derivatives 2d,e (2d:3d=9:1, combined yield 79%; 2e:3e=19:1, combined yield 57%).

Compound 2d was treated with BzONa, BzOLi, t-BuOK or NaH under various reaction conditions, and the mixture of α- and β-anomers (5) obtained after chromatography was analyzed by ^{1}H NMR spectroscopy in each case. Although a BzOLi/DMF system (110°C, 5 h) gave a rather good result (β/α=3/1, combined yield 55%), the other stronger bases caused extensive anomerization.

REFERENCES

1 Sasaki, T., Minamoto, K. and Hattori, K. (1973) J. Org. Chem., 38, 1283.
2 Idem. (1974) Tetrahedron, 30, 2689.
3 Sasaki, T., Minamoto, K., Hattori, K. and Sugiura, T. (1975) J. Carbohydr.·Nucleosides·Nucleotides, 2, 47.
4 After the common dibutylstannylation reaction in MeOH followed by evaporation, the dried residue was treated with roughly stoichiometric amount of TsCl in pyridine[Katsuhiko Yamaguchi: graduation thesis, Faculty of Engineering, Nagoya University (1980)].
5 When R is Tr, tosylation does not occur.
6 Partial deacylation occurred even during the dibutylstannylation reaction in refluxing methanol.

Synthesis and properties of oligodeoxyribonucleotides containing N^6-methoxyadenine

Hisayoshi Nishino, Akira Ono, Akira Matsuda, and Tohru Ueda

Faculty of Pharmaceutical Science, Hokkaido University, Kita-12, Nishi-6, Kita-ku, Sapporo 060, Japan

ABSTRACT

Oligodeoxyribonuclotides containing N^6-methoxyadenine (mo^{6}A), a promutagenic base, were synthesized. The Tm values and thermodynamic parameters for double strand formation of d(CCTGGTAmo^{6}ACAGGTCC) have been measured. The order of the base-pair stability was A-T>mo^{6}A-C$\simeq$mo^{6}A-T>A-C.

INTRODUCTION

O-Methylhydroxylamine is a mutagen in many biological systems[1] like hydroxylamine. The direct substitution of the amino function of poly(C)[2] and adenosine[3] with *O*-methylhydroxylamine has been observed to give the respective *N* -methoxy derivatives. Since mo^{6}A can exists as an amino-imino tautomeric mixture in solution[4] and pairs with U and C[5], it is interesting to measure this type of pair-formation more precisely. We synthesized oligodeoxyribonucleotides containing mo^{6}A and measured the Tm values and other parameters for their base-pair formation.

RESULTS AND DISCUSSION

SYNTHESIS

Oligonucleotides were prepared by the solid-phase procedure using phosphoramidite chemistry on Applied Biosystems 381A. N^6-methoxy-2'-deoxy-adenosine was introduced into the oligomer without protection of the 6-methoxyamino function. The oligomers synthesized were as follows:

d(CCTGGTAXCAGGTCC) X= mo^{6}A, A, G

d(GGACCTGYTACCAGG) Y= A, G, C, T.

Table 1 Thermodynamic parameters

Base pair	ΔG (Kcal mol⁻¹ at 298k)	ΔH (kcal mol⁻¹)	ΔS (kcal mol⁻¹ K⁻¹)
mo^6A-T	- 6.2	- 91.2	- 0.285
mo^6A-C	- 6.4	- 103.1	- 0.324
A-T	- 8.3	- 101.9	- 0.314
A-C	- 2.5	- 56.7	- 0.182

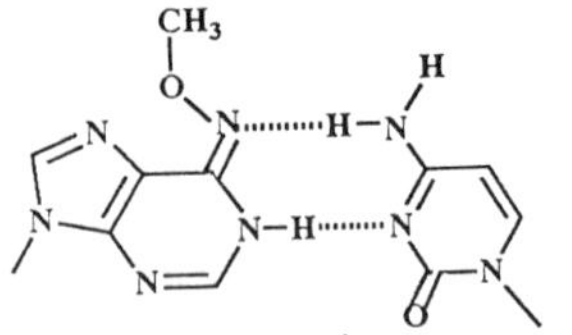
Structure A

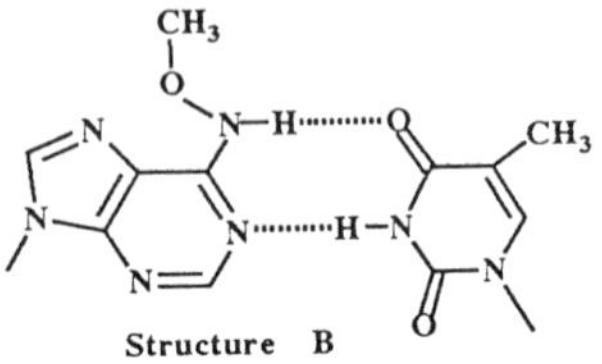
Structure B

STABILITY OF BASE PAIRS

For the Tm measurements equimolar amounts of oligomers were mixed in 0.01 M sodium cacodylate-0.1 M NaCl at pH 7.0. Thermodynamic parameters for the double strand formation were measured for mo^6A-T, mo^6A-C, A-T,and A-C base pairs using the van't Hoff method (Table 1). From Table 1, it is evident that the stabilities of the base pairs follow the order:

A-T > mo^6A-C$\simeq$$mo^6A$-T > A-C.

It is to be noted that mo^6A-C pair is equally stable to the mo^6A-T pair. This may be explained if the imino-tautomer of mo^6A which can be paired with C (Structure A) is as stable in polar solution as the amino-tautomer (Structure B). The ^1H-NMR study of these base-pair formation is in progress.

REFERENCES

1. a) Freese, E. and Freese, E. B. (1966) Rad. Res., suppl.**6**, 97-140. b) Bresler, S. E., Kalinin, V. L. and Perumov, D. A. (1968) Mut. Res., **5**, 1-14. c) Singer, B. and Fraenkel-Conrat, H. (1969) Prog. Nucleic Acid Rea. Mol. Biol., **9**, 1-29. d) Budowsky, E. I., Krivisky, A. S. and Preobrazhenskaya, E. S. (1979) Mut. Res., **59**, 285-289.
2. Fraenkel-Conrat, H. and Singer, B. (1972) Biochim. Biophys. Acta , **262**, 264-268.
3. Budowsky, E. I., Sverdlov, E. D. and Monastyrskaya, G. S. (1971) Biochim. Biophys. Acta ,**246**, 320-328.
4. a) Kierdaszuk, B., Stolarski, R. and Shugar, D. (1984) Acta Biochim. Pol., **31**, 49-64. b) Stolarski, R., Kierdaszuk, B., Hagberg, C-E. and Shugar,D. (1984) Biochemistry ,**23**, 2906-2913. c) Stolarski, R., Kierdaszuk, B., Hagberg, C-E. and Shugar,D. (1987) Biochemistry ,**26**, 4332-4337. d) Fujii, T., Saito, T., Itaya, T., Kizu, K., Kumazawa, Y. and Nakajima, S. (1987) Chem.Pharm.Bull., **35**, 4482-4493.
5. a) Spengler, S. and Singer,B.(1981) J.Mol.Biol., **89**, 255-272. b) Singer,B. and Spengler, S. (1982) FEBS Lett., **139**, 69-75.

Synthesis of the carbocyclic oxetanocin A[1] and its related compounds

Yoshiko Sato, Takahiko Horii, Tokumi Maruyama, and Mikio Honjo

Faculty of Pharmaceutical Sciences, Tokushima Bunri University, Yamashiro-cho, Tokushima 770, Japan

ABSTRACT

Carboxetanocin A (III) and several related compounds including carboxetanocin G (VI), 9-(3,3-bis(hydroxymethyl))adenine (X), -guanine (XI) and 9-cyclobutylguanine were prepared.

(-)-Oxetanocin A, a novel oxetanosyl-N-glycoside isolated from the culture filtrate of *Bacillus magaterium*, has antiviral, antitumor and antibacterial activities.[2)] We have recently reported synthesis of the racemic carboxetanocin A (III) from I via II in 9 steps[3,4)] as well as 9-cyclobutyladenine and 9-(3-hydroxymethylcyclobutyl)adenine[4)]. Compound III showed a comparable level of activity to 9-(arabinosyl)-adenine in the vitro assay of HSV-1 (strain pH) in the *Vero* cells.[5)] We design several related compounds such as VI, X and XI.

Synthesis of VI was achieved by the route shown in the scheme 1. Reaction of II[3)] with 2-amino-4,6-dichloropyrimidine gave IV, which was coupled with the diazonium salt to afford Va. Reduction of Va with zinc powder, followed by successive treatments with triethyl orthoformate and with hydrochloric acid gave VI.

Next, reduction of VII[6)] with lithium aluminum hydride, followed by successive treatments with cyclohexanecarbonyl chloride, with palladium black and with chromium oxide gave VIIIa. Reaction of VIIIa with hydroxylamine yielded VIIIb, which was subjected to catalytic hydrogenation with platinum oxide afforded IX. Reaction of IX with 5-amino-4,6-dichloropyrimidine, followed by conventional treatments afforded X (mp 225 - 227 C). An analogous reaction of IX with 2-amino-4,6-dichloropyrimidine followed by conventional treatments yielded XI (mp >280 C). 9-Cyclobutylguanine (mp >280 C) was similarly

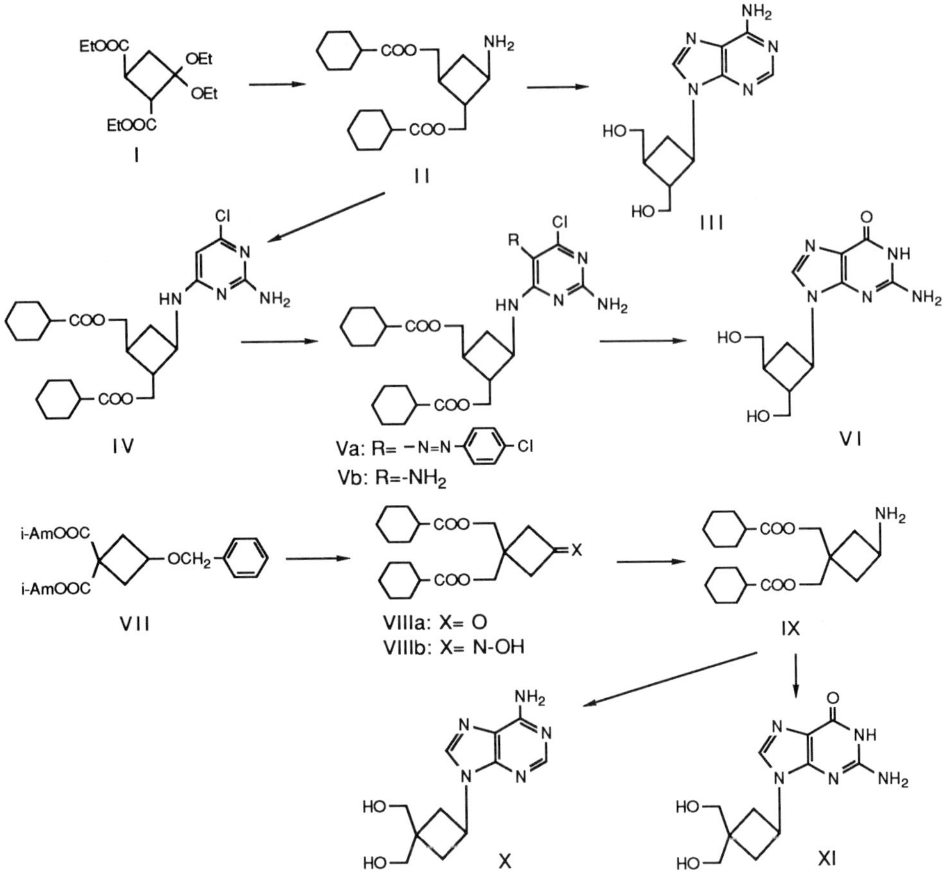

Scheme 1

prepared from cyclobutylamine.

REFERENCES

1. This compound is tentatively named carboxetanocin A.
2. Shimada, N., Hasegawa, S., Harada, T., Tomisawa, T., Fujii, A. and Takita, T. (1986) J. Antibiot., 39, 1623 .
3. Honjo, M., Maruyama, T., Sato, Y. and Horii, T. (1989) Chem. Pharm. Bull., in press .
4. Orally presented at the 109th Annual Meeting of the Pharmaceutical Society of Japan, Nagoya, Apr. 1989.
5. Private communication from Dr. H. Siota, Tokusima University.
6. Wiberg, K. B., Lampman, G. M., Ciula, R. P., Connor, D. S.,Schertler,P. and Lavanish, J. (1965) Tetrahedron, 21, 2749.

Synthetic and high-field NMR study of branched tri-, tetra-, penta-, and heptaribonucleotides modelling the lariat-intron in group II splicing

X.-X.Zhou, J.-M.Vial, A.Sandström, G.Remaud, L.H.Koole[1], and J.Chattopadhyaya*

Department of Bioorganic Chemistry, Box 581, Biomedical Center, University of Uppsala, S-751 23 Uppsala, Sweden and [1]Department of Organic Chemistry, Eindhoven University of Technology, The Netherlands

The lariat structures formed in Group II and nuclear mRNA splicing have adenosine as the "branch-point" residue, linked via a 2´ → 5´ phophodiester bond to a guanosine residue, and a 3´→5´ phosphodiester bond to a pyrimidine residue as shown in the branch-point in **8**, **9**, **18** - **21**. We have recently reported[1] several new routes for the syntheses of branched ribonucleotides **8**, **9**, **18** - **21** and their analogues **1** - **7**, **10** - **17** using a combination of phosphotriester, phosphoroamidite or the H-phosphonate approaches in order to elucidate the conformational and biological significance of the lariat structures in the Group II RNA splicing of pre-mRNA. ^{1}H-NMR conformational studies[2] of six branched triribonucleotides where the branch-point nucleotide was either U, C or G (**1** - **6**), or A in **7** - **13** have been carried out by assigning 270 MHz ^{1}H resonances through 2D NMR and then observing the temperature dependent (i) chemical shifts of the aromatic and the anomeric protons, and (ii) shifts of the equilibrium of N and S pseudorotamer populations of each sugar moiety. The data have been compared with those of U2´ → 5´G, C2´ → 5´G, G2´ → 5´G, A2´ → 5´G and the corresponding diribonucleoside(3´ → 5´)monophohsphates. It emerged that all the branched trimers (**1** - **13**) adopt a conformational state closer to the corresponding 2´ → 5´ dimers than the corresponding 3´ → 5´ dimers. A temperature-dependent ^{31}P chemical shift study[2e,2g] confirmed that the conformational constraint is mainly associated with the 2´ → 5´ phosphate linkage. Although, it appeared from the CD data[2c,2g] that when C or especially when U is at the branch-point the overall constraint is weak. This suggests that even if these trimers adopt a 2´ → 5´ dimer geometry, there is a lack of stabilization by strong stackings within the molecule. This is in sharp contrast with the results found for A in **7** - **13**, and to a smaller extent for G in **5** & **6** at the branch-point. Conformational studies of branched tetraribonucleotides **18** & **19**, on the other hand, showed[2d] overall structures that are reminiscent of a distorted linear 3´ → 5´ linked single stranded A-RNA helix fragment. Conformational studies[2f] with three branched tetraribonucleotide analogues **15**, **16** and **17**, where the branch-point was either cytidine, uridine, or guanosine showed that **15** and **16** are in loose conformational states at room temperature. The possibility for C in **15** and especially U in **16** at the branch-point to be bulged out have been proposed to describe the temperature-dependent ^{1}H chemical shifts. On the other hand, **17** adopts a distorted A-type RNA helix with the distortion at the 3´ terminal U residue instead of the 5´ terminal A residue as in **18**. These studies have shown indications why do all lariat RNA introns have adenosine as the branch-point nucleotide! These conformational studies have furthermore shown that it is the nature of the 5´-nucleotide residue (5´-pyrimidine in **19** versus 5´-purine in **18**) to the branch-point adenosine which dictates the overall conformation due to the intramolecular stackings in the branched RNA[2d]

We have subsequently performed[3] a comparative conformational analysis of tri-, tetra-, penta- and hepta branched ribonucleotides by 600 MHz ^{1}H-NMR and 223 MHz ^{31}P-NMR spectroscopy which have revealed that (i) generation of a lariat structure, typical for Group II and nuclear mRNA splicing, can occur with preservation of the A-geometry of the RNA strand, (ii) mimicry of the lariat branch site requires the presence of at least one nucleotide linked to the 5´-site of the branch-point as in tetramer **18 - 19** or heptamer **21**, since the absence of the 5´-nucleotide, in **9** and **20**, results in an

B (2´p5´B´) (3´p5´B")

1 : B = U, B´ = G, B" = C
2 : B = U, B´ = G, B" = U
3 : B = C, B´ = G, B" = C
4 : B = C, B´ = G, B" = U
5 : B = G, B´ = G, B" = C
6 : B = G, B´ = G, B" = U
7 : B = A, B´ = C, B" = U
8 : B = A, B´ = G, B" = C
9 : B = A, B´ = G, B" = U
10: B = A, B´ = C, B" = G
11: B = A, B´ = U, B" = G
12: B = A, B´ = G, B" = G
13: B = B´ = B" = A

A3´p5´**B** (2´p5´G) (3´p5´U)

15 : B = C
16 : B = U
17 : B = G
18 : B = A

U3´p5´**A** (2´p5´G) (3´p5´C)

19

A (2´p5´G3´p5´U) (3´p5´U3´p5´C)

20

C3´p5´U3´p5´**A** (2´p5´G3´p5´U) (3´p5´U3´p5´C)

21

unnatural molecular conformation which is largely dominated by the 2´ → 5´ stacking. These observations suggest that the formation of the lariat-intron do not seriously alter the secondary structure of the single-strandeed intron and, therefore, may serve as a recognition signal for the final step of the ligation of 3´- and 5´-exons and the excision of the lariat intron.

Acknowledgements

Authors thank the Wallenbergs Stiftelsen for funds for the purchase of a 270 MHz NMR spectrometer, the Swedish Board for Technical Development and the Swedish Natural Science Research Council for generous research grants.

References

1. (a) J.M. Vial, N. Balgobin, G. Remaud, A. Nyilas and J. Chattopadhyaya, *Nucleosides and Nucleotides*, **6**, 209 (1987); (b) G. Remaud, X.-X. Zhou, B. Öberg and J. Chattopadhyaya, in "Reviews of heteroatom Chemistry", vol. 1, ed. S. Oae, MYU publishing Inc., Tokyo, 1988, p. 340-366; (c) X.-X. Zhou, A. Nyilas, G. Remaud and J. Chattopadhyaya, *Tetrahedron*, **43**, 4685 (1987); (d) X.-X. Zhou, G. Remaud and J. Chattopadhyaya, *Tetrahedron*, **44**, 6471 (1988); (e) N. Balgobin, A. Földesi, G. Remaud and J. Chattopadhyaya, *Tetrahedron* **44** (22), 6929 (1988), [f] A. Földesi, N. Balgobin and J. Chattopadhyaya, *Tetrahedron Lett*, **30**, 881 (1989).
2. [a] G. Remaud, J-M. Vial, A. Nyilas, N. Balgobin and J. Chattopadhyaya, *Tetrahedron*, **43**, 947 (1987), [b] J-M. Vial, G. Remaud, N. Balgobin and J. Chattopadhyaya, *Tetrahedron*, **43**, 3997 (1987), [c] L.H. Koole, N. Balgobin, H.M. Buck, W.H.A. Kuijpers, A. Nyilas, G. Remaud and J. Chattopadhyaya, *Recuil des Trav. Chim Pays-bas* , **107**, 515-528 (1988), [d] X-X. Zhou, A. Nyilas, G. Remaud and J. Chattopadhyaya, *Tetrahedron*, **44**, 571 (1988), [e] A. Sandström, G. Remaud, J-M. Vial, X-X. Zhou, A. Nyilas, N. Balgobin, and J. Chattopadhyaya, *J.C.S. Chem. Comm.*, p. 542-544 (1988), [f] G. Remaud, N. Balgobin, C. Glemarec and J. Chattopadhyaya, *Tetrahedron* , **45**, 1537-1548 (1989), [g] G. Remaud, N. Balgobin, A. Sandström, L.H. Koole, A.F. Drake, J-M. Vial, X.-X. Zhou, H.M. Buck and J. Chattopadhyaya, *J. Biochemical & Biophysical Methods*, **18**, 1-36 (1989).
3. L.H. Koole, X-X. Zhou, G. Remaud, H. M. Buck & J. Chattopadhyaya, *J.C.S. Chem. Comm.* (in press)

Synthesis of imidazo[4,5-*d*][1,3]thiazine ribosides and related compounds of biological interest

Chisato Kaneko, Hiroatsu Matsumoto, Sumiko Hara, Tadao Takeuchi, Takeo Mori, Kazuyoshi Ikeda[1], and Yoshihisa Mizuno

Minophargen Research Laboratory, 2-5233, Komatsubara, Zama, Kanagawa 228 and [1]The Center for Instrumental Analysis of Hokkaido University, Kita-ku, Sapporo, Hokkaido 060, Japan

ABSTRACT

Several tri-*O*-acetyl-1-β-*D*-ribofuranosyl- and 2-acetoxyethoxymethylimidazo[4,5-*d*][1,3]thiazine derivatives 5-10 were synthesized from imidazo[4,5-*d*][1,3]thiazine 4 and tetra-*O*-acetyl-β-*D*-ribofuranose or 2-acetoxyethoxymethyl acetate by fusion method. In each case, depending on the substituent on position 5 of thiazine skeleton, only N^3 isomer or both N^1 and N^3 isomers (in different ratios) were obtained.

INTRODUCTION

We have been concerned with studies on the structure-activity relationships of acyclovir and its congeners. As part of this program, we recently reported the synthesis of "acyclic oxanosine" 1 in which N^1 of guanine is replaced with oxygen atom.[1)] In this paper, we report the synthesis of new 5-substituted imidazo[4,5-*d*]-[1,3]thiazine ribosides and acyclo counterparts in order to screen the antiviral activity against herpes simplex type I.

RESULTS AND DISCUSSION

5(4)-Acylamino-4(5)-ethoxycarbonylimidazole 2 was treated with Lawesson's reagent in refluxing xylene to give corresponding cyclized thiazine derivatives 3 in good yeilds. These thiazine compounds were led to the corresponding nucleosides, compound 5 and 6 by fusion method involving fully *O*-acylated-β-*D*-ribofuranose, and also compounds 8 and 9 were prepared from 2-acetoxyethoxymethyl acetate. In the case where R was phenyl group, (namely a-series), a pair of positional isomers 5a and 6a or 8a and 9a were obtained. On the other hand, in the case of methyl substituent, single isomer 5b or 8b was obtained as a predominant product.

The structure assignments were done by ^1H and ^{13}C NMR spectroscopic method and it was found that observed values are in agreement with values reported in other related system.[2)] Compounds 7 and 10 were afforded in moderate yield by $KMnO_4$ oxydation of compounds 5 and 8, respectively.

An antiviral activity of the title compounds will be briefly touched.

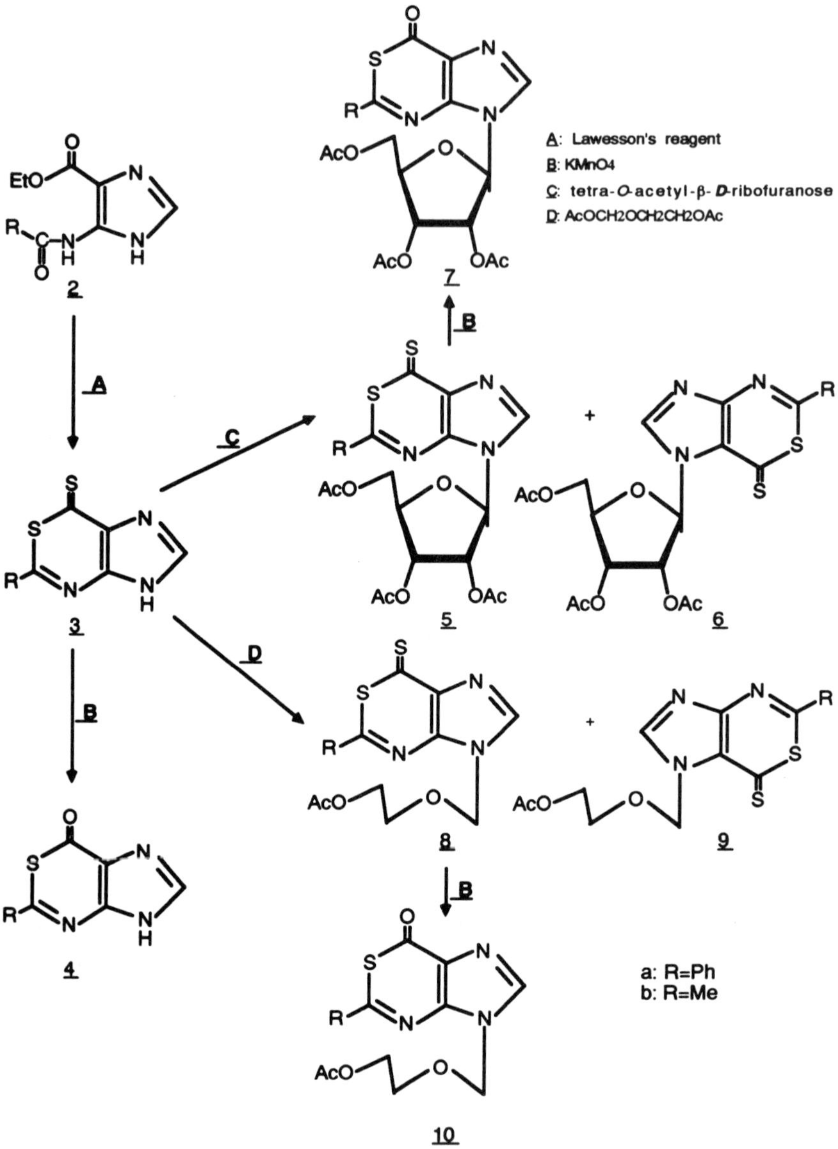

REFERENCES

1. Matsumoto, H., Kaneko, C., Mori, T. and Mizuno, Y. (1989) Chem. Pharm. Bull., 37, 229-232.
2. Poonian, M.S. and McComas, W.W. (1979) J. Med. Chem., 22, 958-966.

Synthesis of novel nucleosides

Masataka Yokoyama*, Naoyuki Yamada, Toshihiro Ikuma, Nobutoshi Obara, Hideo Togo, and Tsuneo Imamoto

Department of Chemistry, Faculty of Science, Chiba University, Yayoi-cho, Chiba City 260, Japan

ABSTRACT

A new spironucleoside structural type and two new bicyclic nucleoside structural types were synthesized from the reaction of ribose nitrile oxide with dimethyl acetylenedicarboxylate. Next, a synthetic method for novel meso-ionic nucleoside structual type is presented.

INTRODUCTION

Many unnatural nucleosides have received considerable attention as antiviral drugs for many viral infections including AIDS. Thus, the purpose of our study is to explore useful synthetic methods for novel nucleoside structual types.

RESULTS AND DISCUSSION

We synthesized a ribose nitrile oxide derivative 2 which was changed in situ into a ribose hydroximolactone 3 at room temperature. Compound 3 was allowed to react with DMAD in the presence of triethylamine to give diastereoisomeric bicyclic nucleosides 5 and 6 in 85 % yield.

The formation of 5 and 6 are considered to proceed through an intermediate, spironucleoside 4a. Thus, we succeeded in obtaining 4b in 20 % yield by successive treatment of 3 with n-BuLi and DMAD in THF at -20 °C.

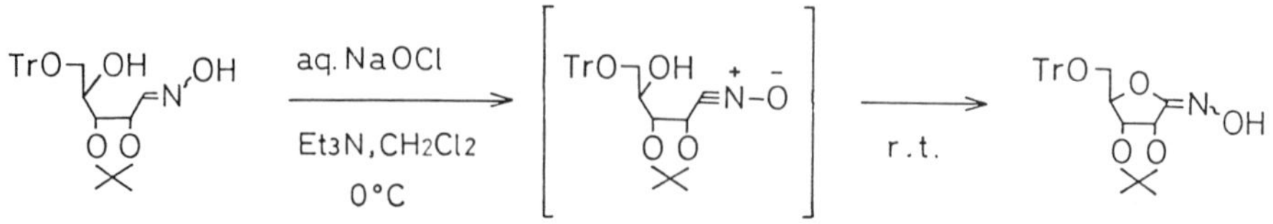

1 2 3

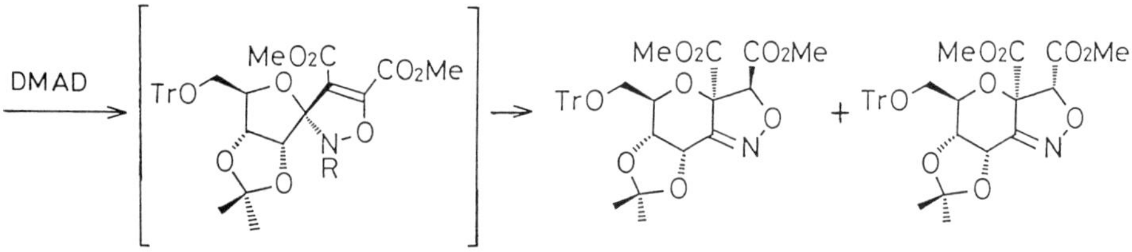

4a: R = ⊖
4b: R = H

5 6

Next, novel meso-ionic nucleoside structural types were prepared by successive treatment of protected deoxyribose 7 with hydrazine, immonium compound, and tetrabutyl ammonium fluoride.

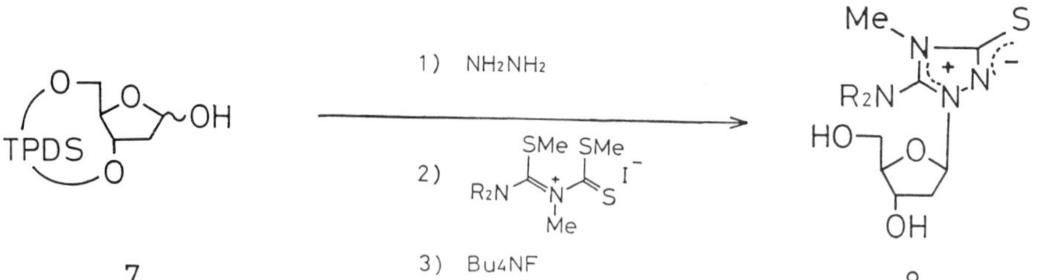

7 8

Effect of pressure on plant endonuclease reactions

S.Kunugi, K.Iwasaki, and A.Nomura

Department of Applied Chemistry and Biotechnology, Fukui University, 910 Fukui, Japan

ABSTRACT

Effect of pressure on plant endonucleases, nuclease P1 from penicillium and an endonuclease from potato, was investigated especially on the influence on phosphomonoesterase and phosphodiesterase activities shown on substrates of XpYp type, as well as their intrinsic pressure-stability. The potato enzyme was found to be far less pressure-sensitive in both senses.

INTRODUCTION

Pressure, another important thermodynamic intensive parameter than temperature, affects enzyme reactions characteristically. In some cases it affects one reaction in a considerably different manner from the other, depending on the nature of the scissile bond of the substrate. Also a certain enzyme show a considerable apparent pressure-activation.

After our studies of pressure effect on various protease reactions we come to think some utilization of these pressure effects on enzyme reactions to control substrate or reaction specificity. One example was recently reported as for protease-catalyzed peptide condensation or synthesis reactions[1]. In such an example the effect pressure is exerted through recognition and reaction by macromolecular catalysts rather sensitive to pressure perturbation and amplifying the effect at lower pressure than what is usually applied in low molecular reactions.

For the reaction concerning with nucleotides, many enzymes show regio-specific catalysis besides strict base specificity and a possibility to control the specificity by external parameter is fairly attractive. Here we take plant endonucleases such as nuclease P1 from penicillium and an endonuclease from potato[2] and investigate the influence of pressure on phosphomonoesterase and phosphodiesterase activity shown on the substrates of XpYp type[3], as well as the intrinsic pressure stability of these enzymes.

EXPERIMENTALS

Potato endonuclease was prepared as reported before and the substrate dinucleotides were obtained by RNAse A and T_1 fragmentation of yeast RNA. Nuclease P1 was purchased from Yamasa Shoyu (Lot D-4,D-7). Reaction products were separated and quantified by HPLC. AMPase activity was followed by amidol-molybdate method. Pressure vessel bearing upto 7000atm (Yamamoto Suiatu Kogyo Co., Hyprex C-7000-19-1) was used for the reaction under high pressure.

RESULTS AND DISCUSSIONS

Table 1 shows pressure dependence of 3'AMPase activity of two nucleases and we see that potato enzyme is far less decelerated by increasing pressure. Such a trend is also seen in the result of dinucleotide hydrolysis as given in Table 2. The notation of the reaction parameters are as in Scheme 1. There is less characteristic difference of pressure effect on two cleavage modes for both enzymes. These results will be related with the mechanistic difference of these two enzymes.

Table 1.
Apparent pressure dependence of AMPase activity of two enzymes.*

Enzyme	pressure (MPa)	relative activity to 0.1MPa.
P_1	200	0.62
	300	0.57
	400	0.34
potato	200	0.94
	300	0.92
	400	0.82

* pH 7.5, 37°C

Table 2.
Pressure dependence of kinetic parameters of dinucleotidase activity.*

Enzyme & subs.	pressure (MPa)	k_{11} (1/h)	k_{12}x100 (1/h)
P_1 GpCp	0.1	1.38	1.75
	200	0.21	0.32
ApUp	0.1	1.27	0.96
	200	0.083	0.076
potato			
GpCp	0.1	0.11	0.59
	200	0.084	0.38

* pH 7.5, 37°C

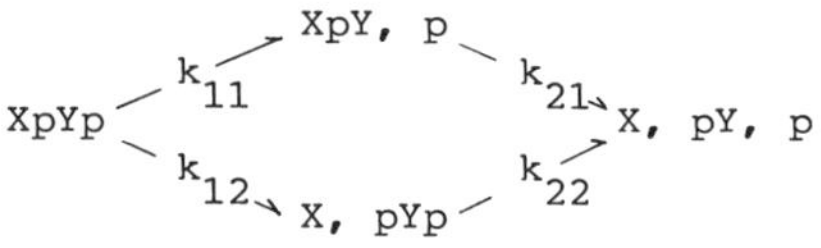

Scheme 1

REFERENCES

1 Kunugi,S. et al. (1987) J.Chem.Soc.,Chem.Comm., 1335-1336;
Kunugi,S. et al. (1989) Bull.Chem.Soc.,Japan, 62, 514-518.
2 Nomura,A.,Suno,M.,Mizuno,Y. (1971) J.Biochem.(Tokyo), 70, 993-1001.
3 Nomura,A. et al. (1976) Nucleic Acid Res. s.p.2, s141-s143.

Symposium Series No. 21 1989 Nucleic Acids Research

Synthesis of DNA fragments bearing 7,8-dihydro-8-oxo-adenine: a gamma radiation product

A.Guy, D.Molko*, and R.Téoule

Département de Recherche Fondamentale, Laboratoires de Chimie, Centre d'Etudes Nucléaires de Grenoble, 85X 38041 Grenoble Cédex, France

ABSTRACT In order to investigate the mutagenic effects and the repair of DNA lesions induced by gamma rays, 7,8-dihydro-8-oxo-adenine was introduced in DNA fragments using cyanoethyl phosphoroamidite method. The modified oligonucleotides were purified by HPLC and the presence of the defect in the final product was confirmed by DNA sequencing, enzymatic hydrolysis and FAB mass spectrometry.

Oligonucleotides bearing a well defined defect in a determined position are useful tools for radiobiologists investigating radiation-induced DNA damage. In the present study, we describe the chemical synthesis of DNA fragments which contain the 7,8-dihydroadenin-8-one moiety.

RESULTS AND DISCUSSION

When DNA is submitted to the action of ionising radiation in aerated solution, 7,8-dihydro-8-oxo-adenine is one of the lesions directly formed (1).

Irradiation cannot conveniently be used to prepare nucleoside 4 (R^1= H) required for the oligonucleotide chemical synthesis. The modified nucleoside 4 (R^1= pac) was prepared from 1 in three steps (Scheme 1).

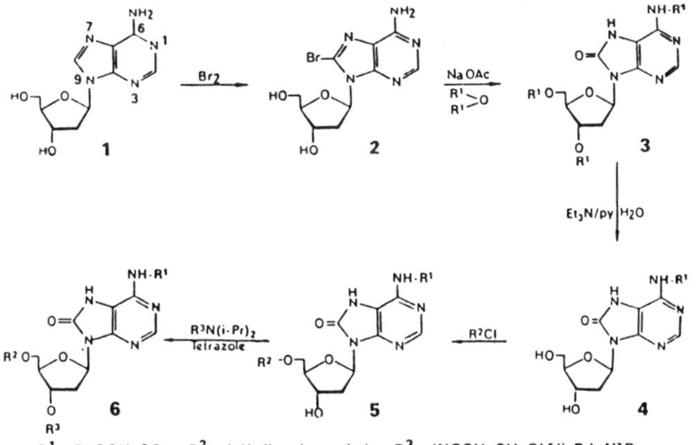

R^1 = $PhOCH_2CO$; R^2 = 4,4'-dimethoxytrityl; R^3 = $(NCCH_2CH_2O)[(i\text{-}Pr)_2N]P$

The method previously described (2) for the bromination of 1 was improved as follows. Compound 2 was esterified and converted into 3 by treatment with anhydrous sodium acetate in the presence of phenoxacetic anhydride and pyridine, thus avoiding N-glycosidic bond rupture. After workup and chromatography, 3 was obtained in 72 % yield. Compound 3 was selectively deprotected to give 4 in 60 % yield. In order to obtain the protected nucleotide required for the chemistry used by synthesizers, 4 was selectively 5'-O protected by a 4,4'-dimethoxytrityl group and further 3' phosphitylation (3) transformed 5 to the phosphoroamidite 6 which was characterized by ^{1}H NMR, ^{31}P NMR and HR FAB mass spectrometry.

A set of oligonucleotides containing the 7,8-dihydroadenin-8-one moiety and of length ranging from 9-mer to 39-mer were synthesized for NMR studies and biological purposes. Each synthesis was carried out on derivatized CPG using the cyanoethoxydiisopropylamino phosphoroamidite method. The oligonucleotides were then deprotected according to a procedure recently described (4).

Modified DNA fragments were purified by anion-exchange HPLC and desalted by dialysis against water. If further purification was necessary, this was done on an ODS reversed-phase column.

To evaluate product syntheses and chain length, the 5'-end DNA fragments were labeled with gamma ^{32}P ATP and polynucleotide kinase and checked by 20 % PAGE. The sequences were confirmed by a modified Maxam-Gilbert method for sequencing oligonucleotides.

The presence of the defect in the final product cannot be directly evidenced by mass spectroscopy under pyrolytic conditions because 7,8-dihydroadenin-8-one and guanine have the same molecular weigth. To verify the nature of the defect in the chain, the DNA fragment was digested with snake venom phosphodiesterase and alkaline phosphatase. To characterize the mixture obtained, a qualitative analysis was performed by HPLC. The peaks were detected by a diode array UV spectrometer and spectra acquired from slope and apex clearly differentiate nucleosides. Product peaks were collected, freeze-dried and checked by MS to be identical to dC, dG, dT, dA and 2'-deoxy-7,8-dihydroadenin-8-one authentic samples.

REFERENCES

1 Téoule R. (1987) Int.J.Radiat.Biol. 51,573.
2 Ikehara M., Vesugi S. & Kaneko M. (1972) Nucleic Acids Chem. 834.
3 Barone A.D., Tang J.Y. & Caruthers M.H. (1984) Nucleic Acids Res. 12,4051.
4 Schulhof J.C., Molko D. & Téoule R. (1988) Nucleic Acids Res. 16,319.

The synthesis and characteristics of adenylyl-(5′→N_ε)-lysyl peptides

B.Juodka and S.Sasnauskiene

Department of Biochemistry and Biophysics, Vilnius University, Vilnius, Lithuania SSR, USSR

ABSTRACT

The dipeptides Boc-Thr-Lys(Cbz)-OMe and Boc-Lys(Cbz)-Glu(OEt)-OEt were prepared by the classical method of peptide synthesis. The Cbz protecting group was subsequently removed via hydrogenolysis. AMP was then covalently joined to the free lysine ε-amine by the carbodiimide (DCC) method. The resulting RNA-ligase active center model compounds, the adenylyl-(5—>N_ε)-lysylpeptides were then used to study the participation of the amino acid residue functional groups in the hydrolysis of the phosphoamide center.

RNA ligase catalyzes ATP-dependent oligonucleotide linkage formation or cyclization in three stages [1]. In the first stage RNA ligase reacts with ATP to form covalent AMP-ligase complexes. Recent work [2] has shown that this occurs through the formation of phosphamide bonds via lysine ε-amino groups. Studies upon the stability of AMP-RNA ligase complexes have determined that adenosine is a decomposition product in aqueous solution. It is known [4] that the phosphamide center of nucleotidyl (P—›N) peptides undergoes bond cleavage only under the conditions where there is a localized hydroxyl or carboxyl group of an amino acid residue. This led to the speculation that the presence of such functional groups in the AMP-RNA ligase complex, localized near the nucleotidylpeptide bond are also required in order for the complex to undergo hydrolysis. Based upon RNA ligase amino acid sequence analysis [5], the lysine residue to which AMP is bound is found between threonine and glutamic acid. The purpose of this work has been to synthesize adenylyl (5'—› N_ε) lysyl dipeptides and to determine the stability of such compounds.

The corresponding dipeptides Boc-Thr-Lys(Cbz)-OMe and Boc-Lys(Cbz)-Glu(OEt)-OEt were synthesized by using the carbodiimide method in solution. The lysine ε-amino groups were then deprotected by hydrogenolysis and the incubation of the corresponding deblocked dipeptides with AMP, in the presence of DCC resulted in the synthesis of adenylyl-(5′—›N_ε) lysinyldipeptide esters (I-II). The treatment of these compounds with strong base gave the free carboxylate analogues (III-IV):

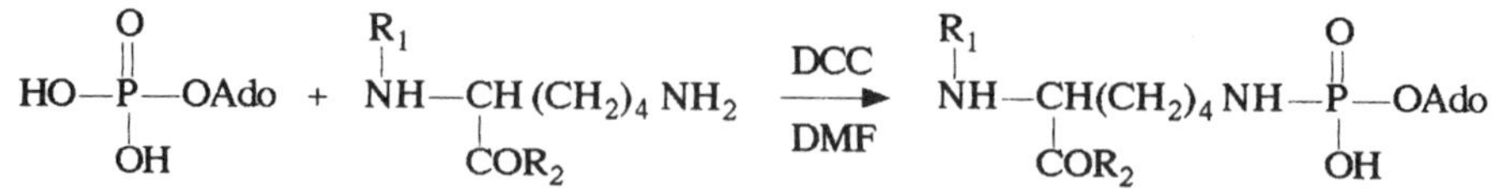

I. Adenylyl-(5′—›N_ε)-(Boc)-L-threonyl-L-lysine methyl ester

R_1 = Boc—NH—CH(CH(OH)CH_3)—CO— ; R_2 = —OCH_3 ;

II. Adenylyl-(5′—›N_ε)-(N_αBoc)-L-lysyl-L-glutamic acid diethyl ester

R_1 = Boc— ; R_2 = —NH—CH($CH_2COOC_2H_5$)—$COOC_2H_5$;

III. Adenylyl-(5′—›N_ε)-(Boc)-L-threonyl-L-lysine

R_1 = Boc—NH—CH(CH(OH)CH_3)—CO— ; R_2 = —OH ;

IV. Adenylyl-(5′—›N_ε)-(N_αBoc)-L-lysyl-L-glutamic acid

R_1 = Boc— ; R_2 = —NH—CH(CH_2COOH)—COOH

Studies upon the hydrolytic characteristics of compounds I-IV determined that all undergo hydrolysis in acidic solution (pH 1-3, 0.01-0.1 M HCl) to give AMP and the respective amino acid dipeptides or their esters. Furthermore, it is of interest to note that both, the adenylyl (5'—›N_ε) lysyl peptide esters and their respective analogues with free hydroxyl and carboxylate groups underwent hydrolysis under these conditions. These results would indicate that the threonine and glutamic acid residues which are located on each side of the the lysine do not participate in the mechanism of nucleotidyl-(P—› N)-dipeptide hydrolysis. It is speculated that these neighboring functional groups are located in a manner that is sterically unfavorable with respect to the phosphamide center. In summary, the results of these experiments would then indicate that the cleavage of AMP-RNA ligase complexes to form adenosine does not depend upon neighboring threonine and glutamic acid functional groups, but rather upon other such amino acid residues, which may approach the phosphoamide center more closely, and which are possibly due to the tertiary structure of the RNA ligase polypeptide.

REFERENCES

1. Uhlenbeck, O.C. and Gumport, R.J., (1981), T4 RNA ligase, In: The Enzymes. New York: Academic Press, **15**, 31-60.
2. Juodka, B.A. and Markuckas, A.J., Snechkute, M.A., Zilinskiene, V.J. and Drigin, Yu. F. (1980) *Bioorgan. Khim.*, **6**, 1733-1734.
3. Juodka, B.A. and Markuckas, A.J., (1985) *Bioorgan. Khim.*, **11**, 821-825.
4. Juodka, B.A. (1984) *Nucleosides and Nucleotides*, **3**, 445-483.
5. Thogersen, H.C., Morris, H.R., Rend, K.N. and Gait, M.J. (1985) *Eur. J. Biochem.*, **147**, 325-329.

Synthesis and selected properties of oligonucleotidyl-($P_m \rightarrow O$)-amino acids

L.Bagdoniene, R.Plaipa, and B.Juodka

Department of Biochemistry and Biophysics, Vilnius University, Vilnius, Lithuania SSR, USSR

ABSTRACT

N-Cbz-Ser(OMe)-(P_m—>O)-d(TpT) was synthesized as a diastereomeric mixture of approx. (1:1) by reacting d(TpT) with 2,4,6-triisopropylbenzenesulfonyl chloride and N-Cbz-Ser-OMe. The phosphoesteric bond of N-Cbz-Ser (OMe)-(P_m—> O)-d(TpT) was found to be stable in acid (1 N HCl, 1 h, 37°C), but labile in alkaline solution (0.1-1 N NaOH, 1 h, 37°C). The products of alkaline hydrolysis were determined to be d(TpT) and the amino acid derivatives. Furthermore, the phosphotriester N-Cbz-Ser (OMe)-(P_m—> O)-d(TpT) was more labile than the diesteric analogue N-Cbz-Ser (OMe)-pdT. The internucleotide phosphotriester linkage of N-Cbz-Ser (OMe)-(P_m—> O)-d(TpT) was also found to be resistant to enzymatic digestion with spleen and snake venom phosphodiesterases.

Recent literature has reported the isolation of nucleic acid-protein complexes from various sources [for review, see (1)]. In most cases it has been shown that the protein and the nucleic acid components are covalently linked via phosphoesteric bond formation between the terminal phosphate of the nucleic acid and hydroxy groups of amino acid residues. Phosphoester bonds may also be formed between the hydroxy group of amino acids and internucleotide phosphates.

The present paper describes the synthesis and properties of the simplest model of the phosphotriester type. For the synthesis of N-Cbz-Ser (OMe)-(P_m—> O)-d(TpT) we have used triisopropylbenzosulphochloride for the activation of the internucleotide phosphates as has been previously described [2].

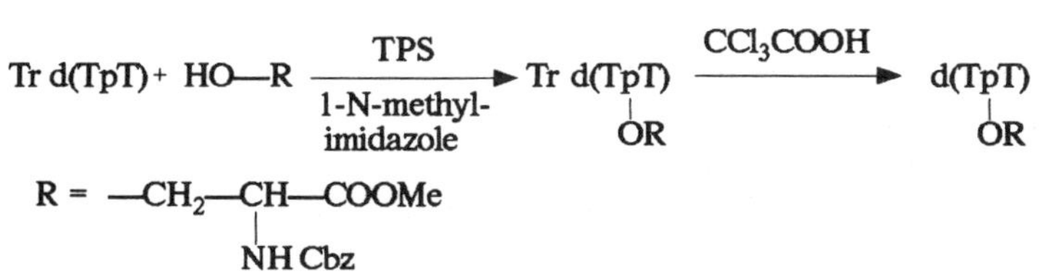

N-Cbz-Ser (OMe)-(P_m—> O)-d(TpT) was given as a diastereomeric mixture of approximately 1:1 on silica gel column, eluted with chloroform, chloroform-ethanol (10%) gradient. The diastereomeric compounds were detectable on TLC (chloroform-methanol (9:1); (Rf 0.34, 0.38, resp.) and by the use of Ultrasphere ODS column, eluted with 30% acetonitrile in water. The structure of the two N-Cbz-Ser (OMe)-(P_m—> O)-d(TpT) diastereomers were also confirmed by H^1 NMR. The molar concentration ratio of thymine:phosphorus:serine of N-Cbz-Ser (OMe)-(P_m—> O)-d(TpT) after complete acid hydrolysis (6 N HCl, 105 °C, 24h) is found to be 2:1:0.9.

The investigation of the hydrolytic stability of N-Cbz-Ser (OMe)-(P_m—> O)-d(TpT) determined that the compound is quite stable in an acidic and neutral medium, but readily undergoes hydrolysis (50-90%) in an alkaline medium (0.1-1 N NaOH, 1 h, 37°C) where d(TpT) is found among the products of hydrolysis. The triesteric compound N-Cbz-Ser (OMe)-(P_m—> O)-d(TpT) was more labile to alkaline hydrolysis than its diesteric analogue N-Cbz-Ser(OMe)-pdT. Since the latter compound has been shown to undergo decomposition via the mechanism of β-elimination [3], it is therefore speculated that the triester would also undergo alkaline hydrolysis by this same mechanism .

Phosphodiesterases are often used for the determination of the presence of phosphodiester linkages in naturally-occuring nucleotidylpeptides [1]. It is unknown whether phosphodiesterases may also be active in the cleavage of phosphotriesters. The experimental results of this work does not support such a contention, as it has been shown that phosphoester linkages of the triester N-Cbz-Ser (OMe)-(P_m—> O)-d(TpT) are resistant to enzymatic digestion with spleen and snake venom phosphodiesterases.

References

1 Juodka, B. (1984) Nucleosides and Nucleotides, 3, 445-483.

2 Jefimov, V.A., Chakmacheva, O.G. (1985) Bioorgan. Chim., 11, 1087-1097.

3 Juodka, B., Kirveliene, V., Pavilionis, P. (1982) Nucleosides and Nucleotides, 1, 111-125.

Author index